PHYSICS AND CHEMISTRY OF BAKING

Third Edition

K. J. DEAN, M.Sc., Ph.D., F.Inst. P., C.Eng., F.I.E.E.
*Principal, South East London College,
Breakspears Building, Lewisham Way, London S.E.4.*

N. E. EDWARDS, M.A.
*Senior Physics Mistress, Ursuline Convent High School,
15, The Downs, London S.W.20.*

C.A. RUSSELL, M.Sc., Ph.D., D.Sc., C.Chem., F.R.I.C.
*Reader in History of Science and Technology, The Open University,
Milton Keynes, MK76AA.*

APPLIED SCIENCE PUBLISHERS LTD
LONDON

APPLIED SCIENCE PUBLISHERS LTD
RIPPLE ROAD, BARKING, ESSEX, ENGLAND

First edition 1963
Second edition 1968
Third edition 1980

British Library Cataloguing in Publication Data

Dean, Kenneth John
  Physics and chemistry of baking.—3rd ed.
  1. Baking
  2. Physics
  3. Chemistry
  I. Title   II. Edwards, Nancy Elizabeth
  III. Russell, Colin Archibald
  IV. Introduction to the physics and
  chemistry of baking
  664′.752   TX763

ISBN 0-85334-867-7

WITH 11 TABLES AND 72 ILLUSTRATIONS
© APPLIED SCIENCE PUBLISHERS LTD 1980

Photosetting by Thomson Press (India) Ltd., New Delhi and
Printed in Great Britain by Galliard (Printers) Ltd., Great Yarmouth

Preface

Since the first edition of this book was written there have been far-reaching changes in the technology of breadmaking. The authors have aimed to incorporate the principles of these in this edition together with changes resulting from the international standardisation of units. This edition has therefore been written in terms of these SI units. Material has been included dealing with activated dough development, the major types of oven now in use and the wrapping of bakery products.

Students in the UK may know that the City and Guilds of London's Breadmaking Technicians Certificate is being phased out and the Technician Education Council's certificate and diploma programmes in the Technology of Food are being introduced. One combination of units in these programmes which colleges offer is intended to meet the needs of technicians in the baking industry. A typical certificate programme of 15 units involves one unit each of physics and chemistry. The text of this book covers those objectives in the TEC standard units Physics I and Chemistry I which the authors consider to be appropriate for this industry, as well as the new material mentioned above.

The authors would like to thank all those in industry and in educational establishments who have been so helpful and who have given their advice so freely.

C.A.R.
N.E.E.
K.J.D.

Contents

Preface v

PART ONE: PHYSICS

1. The States of Matter, Force, Weight and Energy . . 1
2. Relative Density 9
3. Solids, Liquids and Gases 15
4. Viscosity and Surface Tension 22
5. Heat and Temperature 29
6. Measurement of Temperature 34
7. Heat and Temperature Relationship and Specific Heat . 40
8. Changes of State 47
9. Transmission of Heat 57
10. Light and Waves 67
11. Principles of Current Electricity 76

PART TWO: CHEMISTRY

12. Chemical Changes 89
13. Solutions and Colloids 96
14. Atoms and Molecules 105
15. Air: Its Components 122
16. Air: Its Reactions 139
17. Water 150
18. Acids, Bases and Salts—I 160
19. Acids, Bases and Salts—II 172
20. Oxidation 188
21. Basis of Organic Chemistry 200
22. Chemical Basis of Food 207

Appendix: Atomic Weights 219
Answers to Numerical Exercises 220
Index 223

PART ONE

PHYSICS

N. E. Edwards and K. J. Dean

Chapter 1

The States of Matter, Force, Weight and Energy

The physicist divides the world around him into two main classes. These he calls matter and energy.

In the former and more obvious of these, there are those things he can touch and handle; earth, sea and wind for example can all be felt, although perhaps to differing extents. These material things are easily subdivided into three states—solid, liquid and gas. Later it will be seen that the boundaries between these states are not always very clear. It may often be a matter of opinion whether a substance is, say, liquid or solid. Glass and dough, though very different in many ways, both possess many of the properties of solids and some of those of liquids too. How then are the states of matter to be defined? Whilst the majority of substances fit neatly into one of the three states, the possibility of 'awkward customers' must be remembered.

Solids are defined as having a definite shape and size, whereas liquids and gases have no shape of their own, but take the shape of their container. Hence, in a cupful of sugar, the grains of sugar retain their individual shapes as solids, but a cupful of water takes the shape of the cup. Gases and liquids possess one common property, that of flowing. Thus it is that, when it is desired to refer to them collectively, they are frequently called fluids. A fluid, therefore, is a name given to substances that flow. Some fluids flow relatively quickly, such as water or milk, and others flow slowly, such as syrup. The term fluid covers all non-solids.

A gas, as distinct from a liquid, has neither shape nor size. Thus a gas tap left on for a second or two will cause the smell of gas to be noticeable throughout the room. This shows that a gas has no size of its own, but takes the size and shape of the space in which it is confined. The gas from the tap, now in the room, takes the shape of the room, and so also does the air in the room. In a mixture of gases, each has the same size, although the quantity of each present is not necessarily the same.

Under suitable conditions a substance may change from one

state to another. Thus butter melts, and water boils, when heated. The study of matter and, in particular, of changes of state as they are called, will lead also to a study of heat. It is now known that heat itself is not a material thing. 250 years ago it was thought that heat could be wrung from a solid like water from a sponge, until no heat was left. But we now know that heat is an example of the second of the two divisions, of which matter was the first. Heat is a form of energy.

The study of energy necessitates the introduction of new terms and hence new definitions. Two such terms are 'force' and 'work'. If force is defined first then the meanings of work and energy follow. The idea of force is really only a convenient way of explaining certain everyday happenings and their consequences. *Force* may be defined as '*that which causes or attempts to cause motion or a change in motion in any material object*'. This definition is often extended to include the case of outer space, where no force may be acting and bodies would not be still but moving at constant speed in a straight line. The introduction of a force, however, would alter this state of affairs. On earth there is always at least one force acting on every object—the force of gravity—and it causes, or attempts to cause, objects to move towards the earth. This force of attraction between the earth and other objects becomes less as the distance between the earth and any object is increased.

An object, say a 6 kilogram (kg) weight, if moved outside the influence of the earth or of other heavenly bodies, would experience no force on it and, if it were once at rest, it would stay at rest. If it were moving, it would continue to move, and at the same speed. However, the act of taking the weight into outer space has not altered the amount of metal in it. The amount of material in any object is unaltered by such an experiment.

This 'amount of material' is called the 'mass' of the object. The mass of an object is, therefore, constant, and does not alter as the result of being moved from one position to another. Although the mass of the object has not changed by taking it away from the earth, the force of gravity on it has altered. This force is responsible for the weight of the object: it is the weight that has changed. *Weight*, then, is defined as 'the force of gravity acting on the mass of an object.'

Thus, although the amount of metal in the 6 kg weight, that is the mass, is unchanged, one cannot rely on gravity pulling it with a force of exactly 6 kg wt no matter where it is placed. It must be remembered that gravity exerts a variable force, changing slightly over the earth's surface and changing greatly away from it.

If the 6 kg weight is compared with a 12 kg weight having twice

MATTER, FORCE, WEIGHT AND ENERGY

the mass of the same metal, it will weigh twice as much if it is weighed at the same place. Again, three times the mass would weigh three times as much. This may be summarised by saying that the weight of a substance is proportional to its mass. If the 6 kg mass only weighed 1 kg wt on the moon, then the 12 kg mass would weigh 2 kg wt there.

By law, traders' weights must be tested by comparison with standard weights. This is carried out by an inspector of weights and measures and he stamps those he certifies as accurate, to protect both the trader and the public. A weight stamped in, say, London, could be used anywhere in the United Kingdom, and if subsequently taken to anywhere on the earth's surface, the force of gravity acting on it would have altered at the most by not more than about five parts in one thousand.

Unfortunately, some confusion arises because mass is measured in kg, and weight is often expressed in kg wt. Not only are the terms similar but writers sometimes use them without any regard to the important distinction between mass and force that has just been discussed. However, weight is often expressed in newtons (N). The newton is the unit of force used in the international system. The force of gravity acting on a 1 kg mass on earth is 9.8 N. On 2 kg the force is 19.6 N.

Now that force has been defined and weight put forward as an example of a force it is possible to consider the term *work*. Work is said to be done when a force moves a body.

At first sight, this definition may seem illogical, for a man may exert himself all day, pushing against a locked door which does not move as a result of his effort, and yet do no work. In this book, however, the word 'work' will be used only when a force produces motion. The further the distance moved by the force, the greater is the work done. The work done is said to be proportional to the distance moved in the direction of the force. Again, the greater the force required to move an object through any given distance, the greater is the work done. It is said that the work done is proportional to the force employed. In the simplest case, in which all the action takes place in a straight line, it follows that:

$$\text{Work} = \text{distance} \times \text{force}$$

No one force acting alone can cause motion to take place in any other direction than that in which the force acts, so that if a force acting in one direction results in motion in another, in reality more than one force has been acting on a body. Such a case will not be pursued here.

In the simple case, if the distance is measured in metres (m) and the force in newtons (N) then the corresponding unit of work is the joule (J).

This may be illustrated by an example: if a 5 kg weight falls 2 m then the force of gravity has done $5 \times 9.8 \times 2 = 98\,J$ of work. Again, if a man lifts the 5 kg weight through a vertical height of 2 m, he has done 98 J of work against the force of gravity. In the first case the work was done by gravity and in the second the work was done against gravity. It is important to notice the difference between a force doing work and work being done by some other agency against a force, for the second case is really a simple example of a two-force problem in which the man exerts a force in the opposite direction to that in which gravity is acting.

Finally, consider the 5 kg weight held at a fixed height of 2 m above the ground. Gravity could only do work if the weight were free to fall. This potentiality for doing work will be defined as energy. In this present example the weight is said to possess a potential energy of 98 J.

Should the weight be allowed to fall freely under the attraction of gravity, it would lose potential energy as it fell. At a height of one metre, for example, it possesses only $5 \times 9.8 = 49\,J$ of energy. It has lost 49 J of potential energy in falling one metre. The potential energy it has lost by its change of position has been changed into the energy of motion—kinetic energy. At the instant before hitting the ground, all the 98 J potential energy has been changed into 98 J kinetic energy. As an object falls it increases speed. Thus the force of gravity changes the potential energy into kinetic energy by increasing its speed. After hitting the ground, the weight, if it does not rebound, possesses neither potential nor kinetic energy. This energy has gone in the sound of impact and in warming up the ground near it, although only to a very small extent.

In baking machinery we are often concerned, not only with how much work is done, but how quickly it is done. A machine which does one joule of work in one second is said to be working at a rate of 1 watt (W), whilst one which does 1000 joules in one second is said to be working at 1 kilowatt (kW). Commonly one bar of an electric fire is designed to use 1000 joules every second of electric power. Hence it would be called a 1 kW element. If the bar continues to use electric power at this rate for one hour it is said to use a kilowatt-hour of energy. Since there are 3600 seconds in one hour there are 3600×1000 joules of energy used in this hour. The rate at which a machine works (measured in watts or kilowatts) is said to be its power.

MATTER, FORCE, WEIGHT AND ENERGY 5

The work done divided by the time taken gives the rate of working or power. Similarly, the power of a machine multiplied by the time for which it is working gives the work done.

Thus, power in watts × time in seconds = work done in joules.

1 watt-second = 1 joule
1 watt-hour = 3600 joules

In the Chorleywood process of mechanical dough development for breadmaking (CBP), the amount of energy given to the dough by the mixer is expressed not in joules, but in watt-hours. Eleven watt-hours (11 × 3600 joules) have to be given to each kilogram of dough. All this energy has to be given in a very short time, usually under two minutes.

In practice the work obtained from a machine is less than the energy supplied to it due to heat generated in the machine. The ratio of the useful work got out to the energy supplied, expressed as a percentage, is called the efficiency of the machine.

Revision 1.1
1. Write concise definitions of the following terms, with the aid of the text if necessary, and then again without referring to the text: solid; liquid; gas; force; mass; weight; work; energy; power.
2. Suggest definitions for the following: matter; state of matter; length; potential energy; kinetic energy.
3. How many joules of energy are used by a 1 kilowatt fire in 1 minute?
4. If a machine uses 4200 joules in 7 minutes, at what rate is it working?
5. A machine does 5 watt-hours of work. How many joules of work are done?

In discussing mass and weight, no reference has been made to the term *volume*. This is defined as 'the space occupied by a body.' Note particularly that this is not the same as the mass of the body, for the quantity of material in the body, measured perhaps in kilograms, cannot be the same thing as the space which that material occupies, and which may be measured in cubic metres.

It may be noticed, though, that for a particular solid or a liquid, the volume is very nearly a constant quantity for a given mass of that substance.

The volume of a gas is very dependent on other factors, for example, on the forces acting on some enclosed gas, or on the volume of the space into which the gas may be escaping. Here liquids and solids only will be considered.

Two solids of equal mass, lead and aluminium, both free from any imperfections, such as hidden cavities containing air, will also be of equal weight. It is a matter of a common experience that equal weights of lead and aluminium are not contained in equal volumes.

It is now known that all materials contain the same elementary particles which make up the atoms and molecules of which you will be reading in the chemistry chapters. It seems, therefore, that these particles are less densely packed in the substance aluminium than in lead. The denser substance needs less volume for the same mass than the less dense one.

Alternatively, if the volumes of the two substances are equal, the denser would have the greater mass. A cupful of flour weighs less than the same cup when filled with water. We may summarise these observations as follows: (1) the greater the density, the less the volume for a fixed mass; (2) the greater the density, the greater the mass for a fixed volume. This is often expressed in one concise formula:

$$\text{Density} = \text{mass} \div \text{volume}$$

If the mass is measured in kilograms and the volume in cubic metres the density is expressed in kg/m^3: if the volume is expressed in litres (there are 1000 litres in a cubic metre), then the density is in kg/litre. The idea of density is a very important one, since it yields so much more information than is expressed by a knowledge of weight or of volume alone.

100 g of gold and 100 g of wool have vastly different volumes, and a litre of whisky weighs less than a litre of milk. Also, it is of common experience that, when liquids and solids do not mix or dissolve, the less dense substance floats on the more dense liquid.

It is often useful to measure the density of a substance. In some industries this is done as a routine check on bulk materials delivered. The most straightforward way of measuring this is to weigh the substance and to measure, or calculate, its volume. The technique of using a balance will not be considered here, so the main task is the measurement of volume and the application of the results in the formula 'density = mass ÷ volume'. The volume of a regular solid can be calculated from its dimensions, so that it is only with irregular solids that difficulty may arise.

One of the following two methods may be applicable:
(1) Select a liquid with which the solid does not react and in which it does not dissolve, but in which it sinks. Place enough

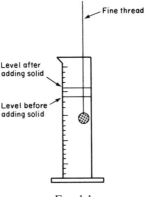

Fig. 1.1

liquid in a graduated jar, or measuring cylinder, so that, when the solid is added later, it can be entirely submerged. Read the volume of the liquid from the scale on the side of the cylinder. Add a known mass of the solid; read the combined volume of the liquid and solid (see Fig. 1.1). The volume of the solid can be found by subtraction, then by dividing the mass by the volume the density can be found.

(2) If a liquid cannot be found in which the solid sinks (this may be so in the case of a solid such as cork), the solid must be attached to a suitable sinker whose volume is found first. The method is then similar to that above.

Revision 1.2
1. Write down the formula for density in terms of mass and volume and compose a short definition of 'density'.
2. Which is more dense in each of the following pairs of substances: Give reasons for your answers.
    (a) milk or cream
    (b) animal fat or water
    (c) a new laid egg or a stale egg?
3. Explain briefly how you would find the density of milk. State carefully how you would find the weight of the milk. If possible carry out the experiment. Should you use (a) as much milk as possible, or (b) as little milk as possible, in order to produce an accurate result: give reasons for your choice. Why may your result not be quite the same as that shown in the table at the end of Chapter 2?

Densities of Some Common Substances

Material	Density expressed in kg/litre
Water	1.0
Butter	0.92
Bone	1.8
Lard	0.93
Gelatine	1.27
Vinegar	1.02
Silver	10.5
Lead	11.4

Chapter 2

Relative Density

In the previous chapter some methods were described for ascertaining the density of a solid. Now let us consider an experiment to determine the density of castor sugar. It would be necessary to find, first the weight, then the volume of the sugar. If the castor sugar were poured into a dry measuring cylinder and carefully shaken until the surface was level, the reading of the surface level in the cylinder might be taken as the volume of the sugar. This is the 'bulk volume' of the sugar, for it is made up of the volume the sugar occupies in the cylinder together with the air spaces between the sugar particles.

When the same weight of granulated or lump sugar is used instead of castor sugar, different volumes are found. This is because the volume of air trapped between the particles is different for each product, and not because of any chemical difference between the three substances.

The bulk volume may be used to calculate the density of the sugar and, since there were three different bulk volumes, there are three corresponding 'bulk densities'. The bulk density of the substance, then, depends on the volume of air trapped between the particles.

To find the solid volume of sugar present in this bulk volume, the sugar must be added to a suitable liquid, in which it does not dissolve. Turpentine is often used for this purpose. Some turpentine is poured into a measuring cylinder and its surface level noted. The sugar is added and the new level recorded. The difference between these two readings is equal to the volume of sugar that has been added. By dividing the mass of sugar by this volume (the solid volume) the solid density is obtained. In an actual experiment, the sugar is weighed, since its weight is proportional to its mass, and its weight used in calculations of density.

Each of the forms of sugar in the above experiment is, in fact, the same substance, and so the solid density will be the same for each. Also the solid density will always be greater than any bulk density, because the volume of trapped air is not considered. Only when there is no trapped air will bulk and solid densities be equal.

It is often useful to know what percentage of trapped air exists in a product. This is clearly important when a commodity is bought or sold by volume alone. A cubic metre of fine gravel may be a better bargain than the same volume of coarse gravel at the same price, since more gravel may well be present in the first case. The percentage of trapped air is often referred to as the percentage of voids in a substance.

The volume of trapped air = bulk volume − solid volume, and percentage voids =

$$\frac{(\text{bulk volume} - \text{solid volume}) \times 100}{\text{bulk volume}} \%$$

$$\text{or} = \frac{\text{bulk density} \times 100}{\text{solid density} - \text{bulk density}} \%$$

Example 2.1

	Icing	Castor	Granulated	Demerara
Weight used (g)	6	6	6	6
Bulk volume (cm^3)†	3.8	4.1	5.8	6.9
Solid volume (cm^3)	3.8	3.9	3.9	4.0
Solid density (g/cm^3)	1.58	1.54	1.54	1.54
Bulk density (g/cm^3)	1.58	1.46	1.03	0.87
% voids	0	5	32	42

† One cubic centimetre (cm^3) is one thousandth of a litre. Density in g/cm^3 is numerically equal to density expressed in kg/litre.

Revision 2.1

1. The above example is the result of an actual series of experiments. Which product would you buy if all sugar were sold by the bag, the bags being of equal size, and the grade of sugar unimportant? Give your reasons. Why is the solid density very nearly the same in each case?
2. Which of the following pairs is likely to take up more room:
    (a) 1 tonne of large coal or 1 tonne of small coal?
    (b) 1 kg of shelled almonds or 1 kg of shelled walnuts?

Density has been described as the mass of unit volume of a substance. The density of stone, for example, is often quoted as

RELATIVE DENSITY

about 2 kg/litre and that of pure water, 1 kg/litre. Water is one of the most common substances and it is often convenient to compare the densities of other materials with that of water.

It can be seen from the figures quoted above that stone is more dense than water, and closer examination shows that stone is twice as dense as water. Thus, the density of stone relative to that of water is two. Alternatively one may say that stone has a *relative density* or specific gravity of 2.

The relative density of a substance is found by dividing its density by that of water, and as such is a numerical constant, not carrying units of weight or volume. The relative density of a substance is a number, although not necessarily a whole number. Because it is a pure number, the relative density of a substance will always be the same, no matter what system of units is used in the measurements for its determination. Comparative figures of the densities and relative densities of some common substances are given at the end of this chapter.

One accurate method of measuring the relative density of a substance is due to the ancient Greek philosopher, Archimedes. He found that when a substance is immersed in a fluid it appears to lose that part of its weight that is equal to the weight of the fluid that has been displaced by the substance. This important statement is known as Archimedes' principle. Note first that the word 'fluid' is used. The principle applies to all fluids, whether liquid or gas. If a weighing were carried out in water by suspending the body from a balance so that it was completely immersed in water, the 'weight' found would be less than in air. This is because of the weight of the displaced water. The apparent weight in any fluid is the true weight less the weight of the displaced fluid.

There are two interesting cases where Archimedes' principle may be applied:

Case 1: If the fluid is water and the object is completely immersed, then, by the principle, it is seen that the loss of weight of the object is equal to the weight of the displaced water. From a knowledge of the density of water, the volume of the displaced water can be found, for the volume of the water is equal to its weight divided by its density. This volume is that of the solid also, for two materials cannot wholly occupy the same space at the same time. Hence the volume of the solid can be found. On our system, pure water, you will remember, has a density of 1 kg/litre (or g/cm^3).

Example 2.2

A solid weighs 78 g in air and in water it weights 68 g

Loss of weight = 78 − 68 = 10 g
Therefore, weight of water
  displaced = 10 g (by Archimedes' principle)
or volume of water displaced = 10 cm^3, since density of pure water is 1 g/cm^3
Hence volume of solid = 10 cm^3 also
Thus density of solid = 78 ÷ 10 (weight ÷ volume)
  = 7.8 g/cm^3
Or the specific gravity of the
  solid = 7.8

This working may be shortened by dividing the weight in air by the loss of weight in water.

$$\text{Relative density of solid} = \frac{\text{weight in air}}{\text{loss of weight in water}}$$

This is an accurate method for finding the relative density of a solid, since it is far easier to weigh with precision than it is to measure a volume accurately.

Case 2: If a solid just floats, completely submerged in a liquid, then it has apparently lost all its weight, and so its weight before immersion is equal to the weight of the liquid displaced by it. Thus the solid must have the same density as the liquid. However, if the object floats only partly immersed, it has lost all its weight in displacing less liquid than would be displaced by all its volume. If the solid is now transferred to a more dense liquid, it will float without displacing so much liquid, and will ride higher out of the liquid.

For the same reasons, in a less dense liquid it will float with more of its volume submerged, or perhaps it will sink.

This second case leads to one possible method for measuring the density of liquids.

Figure 2.1 shows a diagram of an *hydrometer*. The lead shot at the lower end is not heavy enough to sink the instrument completely in the liquids for which it is designed. The chamber containing air assists in keeping the instrument vertical in the liquids. The scale is graduated with the lower densities at the top and the greater ones at the lower end. Hydrometers are usually constructed to be read when floating in transparent liquids, and so the scale is viewed at A, just below the liquid level. Hydrometers are in widespread use: car mechanics use them to test the density of accumulator acid, and H. M. Customs Officers to check the density of gin.

If the liquid is not transparent, and this is often the case with industrial sugar solutions, the instrument is viewed at B but the scale must be calibrated for this type of liquid.

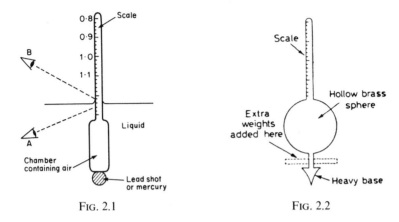

Fig. 2.1   Fig. 2.2

The *saccharometer* is an instrument of this type. It is usually made of brass, and one form is illustrated in Fig. 2.2. In principle it is similar to that of the hydrometer, except that slotted weights are added at the lower end of the instrument to extend its range. When using a saccharometer or hydrometer, a measurement of the temperature of the liquid should be made, as commonly densities fall as temperature rises.

If in the laboratory a density or a specific gravity has to be measured, the temperature should always be stated, together with the result.

Revision 2.2

1. A knob of ice dropped in pure alcohol sinks, whereas in whisky it floats partly out of the liquid. In water it floats yet further out of the liquid. Explain this with reference to the densities of the substances.
2. If the ice in Question 1 is left in the alcohol, after some time it will float entirely submerged, and later float partly out of the liquid. Why is this?
3. The density of a large vat of sugar solution is to be measured by taking a sample for test. Why should the vat be thoroughly stirred first? (this process is called 'rousing the vat'). Place in order of increasing density, a sample (a) from the bottom of the vat before rousing, (b) from the top before rousing and (c) from the roused vat.
4. If an egg just floats submerged in water, how will its position be affected by the addition of brine? Give a reason for your answer.

Common Densities and Relative Densities

Material	Density expressed in $kg/litre$ or g/cm^3
Water (pure; at 4°C)	1.00
Alcohol	0.80
Ice (at 0°C)	0.92
Aluminium	2.7
Copper	8.9
Iron	7.8
Milk	1.03

Chapter 3

Solids, Liquids and Gases

'Solids' and 'liquids'—these terms have been used from the start of this book and are familiar to us all. But can we put into words what we mean by a solid? If you are not able to define a solid, look back to p. 1 where it was defined. The vital property that makes a thing a solid is its ability to keep its own shape and size. But is this a satisfactory definition? A tin can is rolled from flat tinplate, a coin stamped from a copper bar, and a domed roof lined with sheets of lead. The tinplate, the copper, and the lead are all solids, yet they have been made to change their shape. The pressed seam running up the side of the tin can holds the metal in place. If it were undone, the metal would spring back again in an effort to resume its original shape. Copper and lead possess this property to a much less extent than tinplate: spring steel and rubber possess it to a very great extent. This is what is meant by the elasticity of a solid. An elastic substance, when forced out of its original shape and the force removed, tends to go back to its shape again. Elastic is not just a synonym for rubber.

Other substances, dough and putty are examples, are not very elastic. Such substances are called plastic solids. The class of chemicals we know as plastics got their name because they could be forced (sometimes by heat) to take up the shape of a mould and keep this shape indefinitely. In fact, chemical plastics possess very little plasticity at room temperatures after their manufacture and processing is complete, for they can no longer be deformed easily in the same way as dough kneaded by a baker.

Elasticity is an important property of a solid and it must now be considered more closely. Figure 3.1 shows a piece of rubber which has been hung from a stand and which carries markers near its ends and a scale pan and weights at its lower end. The markers move over a scale. It has been found by adding weights to the scale pan that the rubber extends in proportion to the weight added: if the weight is doubled the extension is doubled. It must be remembered that the weight of the scale pan must be included in any calculation

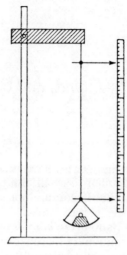

Fig. 3.1

of the weight on the rubber. One must be careful to note that it is the extension that is proportional to the total weight (i.e. the tension in the rubber) and not the total length. An example will show this. A piece of rubber is 60 cm long and, when loaded with 100 g, it is 78 cm long. How long will it be when loaded with 50 g? It is seen that 100 produces an extension of 18 cm, and therefore in proportion 50 g—half as much—will produce an extension of 9 cm. Its total length will then be 69 cm.

Steel is another elastic substance but here the extension produced by a certain weight would be much less than that produced on rubber. Steel can, however, often support a greater load than rubber. But, in both cases, if the load is great enough the point is reached when the stretched wire suddenly becomes thinner (civil engineers who use steel reinforced concrete call this 'necking') and soon snaps. It is important to remember that once the yield point, or limit of elasticity as it is sometimes called, is approached, the extension is no longer proportional to the load for finally a small increase in load can cause the molecules of the elastic substance to slip over each other, and result in a large increase in length prior to breaking.

This law of elasticity has been described as *Hooke's Law*. Hooke's law states that the extension of an elastic solid is proportional to the deforming force (i.e. the load), provided the limit of elasticity is not exceeded.

Hooke's law enables one to estimate the extension of a given

SOLIDS, LIQUIDS AND GASES

sample of an elastic substance once a few initial experiments have been carried out on it. It does not, however, give a direct comparison between the elasticities of different substances. Elasticity is a property of a substance. It should be the same for a thin piece of rubber cord as for a thicker piece of the same material. The extension of the two cords would be different for the same load. Since the load pulls on all the molecules of the cord, if the number of these is increased by, say, four, then the extension is shared between four times as many molecules and the extension is divided by four. To increase the number of molecules by four, i.e. to increase the cross-sectional area by four, the thickness must be doubled.

Again the extension produced by a long cord would be greater than that produced by a short one although each was the same thickness and carried the same weight.

To compare elasticities, then, we must compare extensions of a given length of cord and must compare the forces acting on a given number of molecules, that is on the same cross-sectional areas of cords. Still within the elastic limit, this may be summarised as follows:

1. Extension is proportional to original length.
2. Extension is proportional to force.
3. Extension is inversely proportional to cross-sectional area.

Altogether extension is proportional to

$$\frac{\text{original length} \times \text{force}}{\text{cross-sectional area of cord}}$$

or

$$\text{Extension} \times \text{elasticity} = \frac{\text{original length} \times \text{force}}{\text{cross-sectional area of cord}}$$

$$\text{Therefore elasticity} = \frac{\text{original length} \times \text{force}}{\text{cross-sectional area} \times \text{extension}}$$

If it is borne in mind that the tensile stress is the name given to the force divided by the cross-sectional area then:

$$\text{elasticity} = \frac{\text{tensile stress}}{\text{fractional increase in length}}$$

During dough-making the consistency of the dough changes. The fermentation of the dough causes the elasticity (or if you prefer it, the plasticity) of the dough to change. The dough is said to become softer, i.e., it becomes more plastic and less elastic.

Revision 3.1

1. A 50 cm rubber cord is found to be 80 cm long under a load of 150 g. What will its length be under a load of 125 g? What load causes it to be 55 cm long?
2. In Fig. 3.1 markers were used. What is their purpose? How does their use increase accuracy?
3. A steel rod with round cross-section $\frac{2}{11}$ cm in diameter is subject to a load of 2000 kg. What is the tensile stress in the rod? (Take π as 22/7 and give answer in kg/cm^2.)
4. Find the load in kilograms required to produce a tensile stress of 900 N/cm^2 in a piece of rubber whose cross-sectional area is 0.109 cm^2 (1 kilogram is equivalent to 9.8 newtons).

It can now be seen that a solid may not keep its own shape under the action of a force, and that elastic solids are those that tend to regain their shape when the force is removed. Plastic solids do not do so to any appreciable extent. Also the extent of the elasticity and of the plasticity varies from one solid to another. Now how is plasticity made obvious? By the fact that when the solid has experienced a force, it is deformed in shape. So it seems that a very plastic substance only requires a very small force to alter its shape. Such a force may be found due to the weight of the solid itself. Although a sheet of road-mender's pitch may fall with a metallic noise on the road, if it is laid on the edge of the kerb it will soon have buckled out of shape. A thick slice of butter laid across a pencil, lying on a desk, will soon curve under its own weight. These examples show that plasticity is not far removed from the liquid state, for the butter and the tar, on warming, both distort more easily until finally we say they are liquids.

Liquids have the property of flowing—they take the shape of their container. But do they do so quickly? Water does so very quickly; stock syrup not so quickly, treacle even less quickly, whilst even sheets of pitch in a bucket will eventually take up the shape of the bucket, flowing very slowly and the process taking some years. Yet all are liquids and flow. Those that flow quickly are called mobile and those that flow slowly are said to be viscous.

Just as all solids possess some elasticity (some have very little of it), so all liquids have some viscosity (again some liquids are not very viscous).

The border between solids and liquids is not as easily defined as may be at first supposed. Is pitch a plastic solid or a very viscous liquid? Pitch possesses the property of 'flow' and is, therefore, a liquid, but at room temperatures it flows extremely slowly so that casual observation would not reveal it—otherwise a road surface of pitch or bitumen would end up in the gutter! The measurement of viscosity is discussed in the next chapter, but it can be said here that both the viscosity of pitch and its plasticity can be measured.

Viscosity does not only occur in liquids, it occurs in gases too. When a gas flows in a tube, forces act between the gas and the tube in which it is flowing and between layers of gas itself. In fact viscosity occurs in all fluids, and the term *fluid* includes both liquids and gases. However, gases have some properties which are quite distinct from those of liquids and solids. One of these is the effect of pressure on a gas.

The term *pressure* is one used loosely in everyday life, but more precisely it means the force which is acting on unit area. For instance, if a sack of flour of mass 100 kg and therefore of weight 980 N is transferred to a box of base 2 m × 1 m and spread evenly, then 980 N would be acting on 2 m^2. The pressure due to the flour (a solid), that is the force per unit area, would therefore be 490 N/m^2.

Liquids and gases also exert pressures. For example, the air of the atmosphere exerts a pressure on the earth of 100 000 newtons per square metre. This is because the molecules of which the air is composed are constantly in motion, bombarding the earth's surface.

If gases are enclosed in containers it is easy to exert extra pressure on them by means of a piston and a non-return valve, as in a tyre foot-pump. Alternatively the pump can be motor driven. A dial gauge can be devised to record the pressure. For most enclosed gases at room temperature, halving the volume as the piston moves down the cylinder will double the pressure. When the volume has been reduced to a quarter the pressure will be four times as great as at the start. Finally, in the case of the foot pump, the high pressure air is expelled into the tyre itself. When pressure is applied to a liquid the reduction in volume is extremely small.

If a gas is sufficiently cooled it can be changed to become a liquid. Provided that a gas is not cooled to near its liquefaction temperature then it will behave in the way just described. Gases such as oxygen, nitrogen and hydrogen need considerable cooling before they liquefy so for practical purposes they can be said to act like this. Such gases are said to obey Boyle's Law which states that for a fixed mass of gas at constant temperature

$$\text{pressure} \times \text{volume} = \text{constant}$$

The effect of temperature on the volume of a gas is discussed in Chapter 5.

A certain amount of air is always present in water and most gases are soluble to a slight extent in water, although their solubility decreases as the temperature rises. It is important to distinguish between air dissolved in water and air which is actually beaten into liquids like egg-white to increase their volume and so make them seem light. It is also important to distinguish between gases which merely dissolve in liquids and those which react chemically with them. Water will absorb large quantities of ammonia, sulphur dioxide and chlorine and can never be used for a confining liquid for these gases. Its dissolving power for carbon dioxide is also considerable. When yeast ferments, carbon dioxide is given off and when the amount of carbon dioxide evolved is greater than the water (or milk) in which the yeast is fermenting can hold, then it begins to make froth.

All matter consists of atoms and molecules as explained in Chapter 1. When the number of molecules per unit volume of a gas varies within the gas then some of the molecules will move in order to equalise the distribution. This process is called molecular diffusion. Diffusion also occurs when a space is occupied by two or more different gases originally separated in different parts of the space. As soon as the molecules of the gases are free to do so, they move so that finally the molecules of the gases are evenly distributed. Diffusion can also take place when the gases are separated by some porous material. Commonly unglazed earthenware is cited as a typical material through which molecules can pass easily, but some wrapping materials for bread are also porous.

Waxed paper has been used in England and Wales for many years for wrapping bread, though its use has recently declined, whilst still widely used in Scotland. Packaging of bread is not common on the Continent. New materials have come in which allow less diffiusion and which are suitable for automated wrapping. These include *low density polythene film*, similar to the type used for plastic carrier bags. It is cheap and can easily be inflated on the packaging machine and it does not pick up much static electricity by friction (see Chapter 11). *High density polythene film* is also used which can be much thinner and tougher than the low density film. It is however susceptible to being cut by the sharp edges of crust. Since these films are less permeable than waxed paper they enable the bread to have a longer shelf life. However, mould is more likely to form unless some steps are taken to prevent it.

Revision 3.2

1. List the following substances as: (i) elastic solids; (ii) plastic solids; (iii) viscous liquids; or (iv) mobile liquids. Milk, spring

steel, golden syrup, motor oil, lead, phosphor bronze, waterglass, chewing-gum, cooking fat, bun wash, methylated spirit.
2. What is the effect on the viscosity of a liquid of heating the liquid?
3. Distinguish between viscous and mobile liquids basing your definition on the ease with which they can be stirred.
4. A bubble of air is released from the bottom of a beaker of liquid. If the liquid is viscous does it rise quickly or slowly?
5. Viscosity has been explained as a type of liquid friction, the molecules slipping over each other. Different types of molecules are supposed to have varying degrees of 'roughness'. Use this idea to distinguish between viscous and mobile liquids and to explain why treacle becomes less viscous when mixed with water.
6. A box in the shape of a cube has sides 0.5 m long. If the box contains a mass of 120 kg what pressure does it exert on the floor?
7. A certain mass of enclosed air occupies 300 cm^3 at atmospheric pressure. What volume does it occupy when the pressure is (a) trebled, and (b) halved, by using a movable piston?

Chapter 4

Viscosity and Surface Tension

In the last chapter it was seen that some liquids flow easily—these are mobile liquids. Some flow more slowly—these are said to be viscous. It is often important to be able to measure the viscosity of some liquid, to be able to compare it with the viscosity of some other liquid or to control the viscosity of a liquid. If, for example, a bun wash were too viscous, not only would it not flow easily over the surface of the bread but the baker would almost certainly use too much of it. Think of a bun wash with the viscosity of golden syrup and you will realise that the high viscosity would make it impossible to use.

Again, a chocolate intended for a biscuit manufacturer must have a sufficiently high viscosity or it will run off the biscuit and the coating will be thin. If the viscosity is too high, the coatings will be too thick, the drag of the chocolate in the coating machinery may prevent it working and, in any case, the manufacturer's costs are likely to rise, since he will be using more chocolate.

In the first example you probably realised that the viscosity of the bun wash could be decreased by dilution with a mobile liquid—water. However, in the second case, it is necessary for the chocolate to have a low viscosity during coating of the biscuits, and to have an extremely high viscosity when the biscuits are packed and sold. In this case, then, viscosity is controlled by temperature. The viscosity of a liquid generally falls as its temperature is raised.

These are the methods for controlling viscosity:
 (a) The addition of other less viscous liquids called fluxes or alternatively by adding more viscous ones.
 (b) The control of temperature.

These two points must be taken into consideration in the measurement of viscosity.

It may not be sufficient to say that a liquid is mobile or viscous any more than it is to say that a solid is plastic or elastic or an oven is hot or cold. Such terms are often too vague.

One of the simplest practical ways to measure the viscosity of a

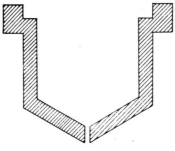

Fig. 4.1

liquid is to use a flow cup. Figure 4.1 shows a section through a simple flow cup. This type is small enough to hold in the hand or it may be supported on a ring clamp and stand. It is made of brass and has a small hole drilled in the centre of its conical bottom. The size of the cup and of the hole and the length of the hole are carefully laid down by the British Standards Institution. It is filled whilst the hole is blocked by a finger under the bottom of the hole. The finger is removed and the liquid allowed to run out. A stop-watch is used to record the time the liquid is flowing, the watch being stopped when the thread of liquid first breaks into drops.

More elaborate viscometers have a water or oil bath around them to keep the liquid being tested at a suitable temperature and have levelling screws and a stopper for the hole which can quickly be removed. In every case a thermometer is used, for it is important to know the temperature of the liquid when the viscosity is measured.

It is usual to quote the viscosity as the time in seconds for the liquid to run out or for a known volume of liquid to run out. Also, since there are a number of different viscometers on the market the name of the viscometer is also quoted. Thus a liquid may have a viscosity of 25 standard flow cup seconds indicating that it took 25 seconds for the liquid to flow from a standard flow cup.

The flow cup is used to test samples of liquid so that the viscosity of the liquid can be held constant, perhaps by the addition of suitable fluxes or adjustment of the temperature of the liquid. When the temperature is much above that of the surroundings, a bath-type viscometer, such as the Redwood viscometer, is used. Figure 4.2 shows the main refinements of this type of viscometer.

Other viscometers depend on the drag on a disc or a cylinder which is rotated when submerged in the liquid. If a cylinder is rotated at constant speed in a liquid, the liquid will exert a viscous drag on the immersed surfaces of the cylinder and force must be used to keep the cylinder rotating.

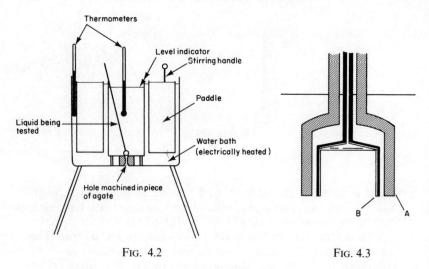

Fig. 4.2   Fig. 4.3

One instrument of this type uses two cylinders and a section through them is shown in Fig. 4.3. One of these cylinders, A, is driven by a motor at a fixed speed, and the other, B, is pulled round only by the viscous drag of the liquid between them. The cylinder B turns against a spring and moves a pointer over a dial. The movement of the pointer depends on the viscous drag of the liquid between A and B and this is a measure of the viscosity of the liquid. Unfortunately, for many liquids, their viscosity alters with the rate at which the cylinder is turning. For example, it needs more effort to stir paint slowly than it does to do so quickly. If a viscometer of this type is used, the speed of the cylinder surface through the liquid must be known.

Revision 4.1
1. How can the viscosity of a bath of liquid be increased?
2. Why is a flow cup not suitable for measuring the viscosity of chocolate? What is the drawback of using it for any liquid when held in the hand?
3. Why is it necessary to record a liquid's temperature when measuring its viscosity?
4. What would be the effect of icing a cake if the icing had (a) too low a viscosity; (b) too high a viscosity?
5. In a cylinder type viscometer, why must the cylinder be rotated at a constant speed?

VISCOSITY AND SURFACE TENSION

It is not easy to find the viscosity of certain animal and vegetable fats at room temperature since their viscosity is so high that they flow extremely slowly and it would require a great deal of effort to rotate a cylinder in them. In fact, if a cylinder were used, air would probably be introduced and the cylinder would slide over the surface of the comparatively hard fat. These substances are better treated as plastic solids. Their plasticity can be found using a penetrometer such as that shown in Fig. 4.4. The brass cone has a hardened tip and is held in contact with the surface of the solid.

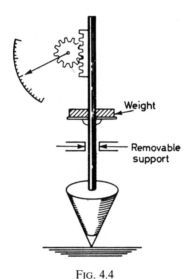

FIG. 4.4

The cone has weights added to it, so that the total load is, say, 100 g. It is then released and allowed to rest on the surface of the solid for 5 seconds (s) before it is again held. The movement of the cone in this time is recorded on a dial and a specimen result might be expressed as a penetration of 0.02 cm, under 100 g, for 5 s.

This method with suitable modifications can be used for substances such as butter, paraffin wax, and even soft road tars.

SURFACE TENSION

In Chapter 14 the elementary ideas of atomic theory are discussed. The idea that all matter consists of small particles, atoms or molecules, leads to the further idea that attractive forces exist between these particles. In the case of a solid, these forces are large and, even if an attempt is made to deform the solid, it tends to resume its original shape. Thus the phenomenon of elasticity can be explained in terms of the molecular nature of matter. In the case of a liquid, however, the binding forces are somewhat weaker, resulting in that characteristic of liquids that is called 'flow'.

At the surface of a liquid a molecule is being attracted back into the liquid by the inter-molecular forces, whereas deep in the body of the liquid one could not easily say in what direction the forces are acting. Therefore, it is to be expected that at the surface of a liquid, the molecules that form the surface will not easily leave it and escape into the atmosphere. This effect is noticeable and is called surface tension.

It is often stated that the surface of a liquid acts as a skin which is in a state of tension, preventing molecules from escaping. It is worth keeping this idea in mind, provided it is also remembered that the liquid itself is the skin, so that this skin is no ordinary one that can be peeled off leaving the molecules free to escape. The skin is always there, for there are always molecules below the surface, keeping the surface molecules from escaping.

Of course, a few molecules do escape, and this process is called evaporation. If the liquid is heated sufficiently, enough energy is given to the molecules to allow them to overcome this force and the liquid is said to be boiling. This is discussed in more detail in Chapter 8.

To learn more about the phenomenon of surface tension, it is necessary to examine the surface of a liquid, say water, very closely. If the water is in a clean glass jar it will be seen that the surface curves upwards near the edge of the vessel. This is shown enlarged in Fig. 4.5. If the wall is quite clean, the liquid surface curves round until it ends at the point A in the figure in a vertical direction. There-

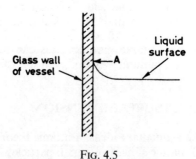

Fig. 4.5

fore, the surface tension is seen to pull upwards all round the edge of the glass vessel. If the glass vessel is a very fine tube then this upwards pull will cause the liquid to rise in the tube (see Fig. 4.6) until the weight of the liquid supported is equal to the pull of the surface on the glass tube. In the case of water this pull is equivalent to a force of 0.07 N for every metre of the perimeter of the edge where the surface tension is acting.

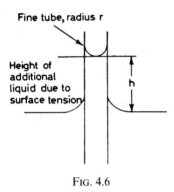

FIG. 4.6

For example, if the tubing has an inside diameter of 2 mm then the total pull is equal to the force of gravity on a mass of

$$\frac{22}{7} \times \frac{2}{1000} \times 0.07 \, \text{N}$$

Remembering that the density of water, ρ, is 1 kg/litre or 1000 kg/m^3, we can find its height.

If the surface tension is T then the volume of water is $\pi r^2 h$, $\pi r^2 h \rho$ must be its mass and $9.8\pi r^2 \rho$ its weight. This must equal $2\pi r T$.

$$\therefore 2T = 9.8 r h \rho$$

where h is the height. In the above example the height is about 1.4 cm.

Now it can be seen that, if the radius (r) of the tube is divided by, say, one hundred, then the height reached is one hundred times as great, i.e., 140 cm high! The finer the hair-like cracks, the more easily the liquid rises. From the Greek word for a hair this effect gets its name capillary action. This causes water to rise in the fine spaces between the fibres of wood or a biscuit to become quickly sodden when dipped in tea.

Capillary action causes water to rise in the cracks in the fired bricks used in house building and damage is prevented in the house by the use of a damp-proof course that is impervious to water. Carpenters insert special grooves or 'throats' when joining together pieces of wood that have to be exposed to the rain, for the finer the workmanship, the closer the join, and the more easily the water can rise. It can only be prevented by the wider throats hidden in the woodwork.

Surface tension also holds the hairs of a brush together as it is pulled out of water. It is responsible for the shape of a soap bubble and for the shapes of water droplets on a dusty floor.

The pull of surface tension can be greatly reduced by the slightest trace of soap or detergent. This is how soapy water thoroughly wets a garment and enables the soap to get at the particles of dirt. If, in the previous example of the capillary tube, the slightest trace of soap were introduced into the water the height of the water column would have been quickly reduced, for the surface tension of the soap solution is much less than that of pure water.

Small pieces of wood with a piece of camphor wedged in one side and floated on water move about on the surface of water due to the changing surface tension as the camphor dissolves. These toy boats are propelled by the unequal forces of the surface tension of the water and of the water close to the piece of camphor.

Baking tins have sloping sides so that liquid held in them is not pulled up so much at the edges by surface tension. Fat solidified in these tins has a flat surface.

Revision 4.2
1. What do you understand by the meniscus of a liquid? What causes it?
2. If the surface tension of a liquid is 0.07 N/m calculate the force in newtons acting on a plate 1 m long which dips into a liquid. (Ignore the thickness of the plate).
3. Account for the practice of laying spacers between the slates of a roof.
4. Water is added drop by drop to a jug until it will contain no more. Account for the appearance of the surface.
5. If water will rise to a height of 2.8 m in a brick wall, estimate the diameter of the cracks in the bricks.

Chapter 5

Heat and Temperature

When a body is heated, the energy given to it causes the molecules of which it is composed to vibrate more rapidly. When this happens (when, for example, a kettle is heated), we say that the *temperature* of the body rises. The temperature of a body is defined as 'its degree of hotness' and any rise in temperature can be measured on a thermometer. This will be further considered in Chapter 6. However, a rise in temperature is not the only result of the heating. Bodies expand as the result of heating. When the kettle of water boils, some of the water changes into steam; eventually all of the water would change if heating were continued. This is an example where heat changes the state of a substance. Again, heat applied to candle wax causes it to melt. Yet a fourth effect can be seen when a piece of metal is heated strongly; the metal turns 'red hot' and eventually 'white hot'.

Heat, therefore, can affect a body in at least four different ways. It can cause a change in temperature, in volume, in the state of the substance, and in its colour. If, in the above examples, the body is allowed to cool, however, it will revert to its original condition as regards its temperature, state, colour and volume, although not always to the same shape, for melted candle wax will, while a liquid, take the shape of the containing vessel and so keep it on solidification.

When the effects of heating are such that the body reverts to its original condition on cooling, the change is said to be an example of a 'physical change'. It should not be confused with a 'chemical change' where the body undergoes a complete change in chemical composition. For example, when bread and cakes are browned they do not lose their colour on cooling, as the heat in this case has caused a chemical change. This will be considered further in Chapter 12.

Not all baking processes involve chemical changes. Meringues are merely 'dried out': that is, the moisture in them is driven off and, if it could be collected, it would be seen to change back to water again.

The effect of heat on temperature and state will be dealt with further in Chapters 6 and 8. The effect of heat on the size of bodies can be demonstrated by a series of experiments.

To show that solids expand on heating, a metal ball is used which will just pass through a metal ring. The ball is then heated and placed on the ring. It will no longer pass through the ring, but will do so again on cooling. This shows that the ball expands on heating and contracts on cooling.

To show that a liquid expands on heating, a flask is fitted with a cork, through which projects a glass tube. The flask is completely filled with the liquid under investigation, and liquid partly fills the tube, as in Fig. 5.1. The flask is now heated. At first the level of liquid in the tube drops slightly due to the expansion of the flask, but after a while it begins to rise rapidly, showing that the liquid expands on heating.

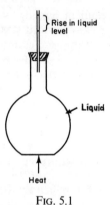

Fig. 5.1

The same apparatus can be used to show that air expands on heating.

The flask is emptied of the liquid and is thus full of air. The flask is inverted, so that the tube dips well under the surface of some water in a beaker. The flask is heated gently and bubbles of air can be seen rising in the water from the bottom of the tube. This shows that the air has expanded and in consequence some of it is being driven out. When the flask is allowed to cool, the air filling it when hot, contracts, and draws water up into the flask (see Fig. 5.2).

Allowance has to be made for the expansion of solids, liquids, and gases in everyday life. Railway lines become longer in summer than in winter and thus gaps have to be left between the ends of

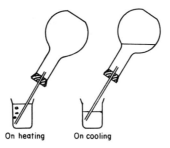

On heating   On cooling

Fig. 5.2

relatively short distances of line to allow for expansion. Modern high speed track still includes a form of expansion gap. Telegraph wires are allowed to sag to prevent them snapping when they contract in cold weather. A bend is made in water pipes to allow for expansion. Allowance has to be made when fitting oven shelves, otherwise they could not be pulled out when the oven is hot. Oven doors are made to fit outside the oven wall so that they do not jam on heating. A small space must be left above the liquid in a bottle when it is filled, if it is to be tightly stoppered, otherwise the contents might crack the bottle if there is a rise in temperature; the expanded liquid will exert a force on the bottle and its stopper.

Expansion is not only a drawback, it is of use too. In the manufacture of beer barrels, rings are put on the barrels when they are hot, so they will hold the wood tightly in place on cooling. Chocolate easter eggs are poured into moulds hot, and when cool, they come away easily from the sides. By far the greatest use for the expansion of liquids is in thermometers for the measurement of temperature. This is so large a subject that it will be dealt with in a separate chapter.

All solids do not, of course, expand to the same extent on heating. To demonstrate this, strips of copper and iron of identical dimensions are riveted together to form a 'compound bar' or 'bimetallic strip' (see Fig. 5.3). The bar is heated and is seen to bend with the copper strip on the outside. This takes place because the copper

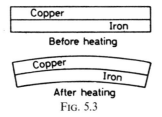

Fig. 5.3

strip has become longer than the iron one, and in the same way as a boat crew takes the inside of a bend in the river to row a shorter distance, the iron strip, which is shorter, is on the inside of the bend. The bar has bent in order that the copper strip may be longer than the iron one, since the rivets prevent them from separating.

The bending of a bimetallic strip is of great use in the thermostatic control of electric irons and ovens. The strip forms part of the electrical circuit as shown in Fig. 5.4. Contact is made between a screw and one end of the strip. The other end is wired permanently to the heating coil. When the temperature rises, the strip begins to bend, and the contact is broken between it and the screw, thus preventing the current from flowing through the coil. As soon as the temperature drops again the contact is re-made and the iron or oven re-heated. The screw can be adjusted so that the contact is broken at the required temperature, and this is the function of the regulator dial on electric irons and ovens. A light bulb is often incorporated in the circuit to indicate when the current is flowing. Bimetallic strips are also used in a pyrometer (see Chapter 6) for temperature measurement (Fig. 6.4).

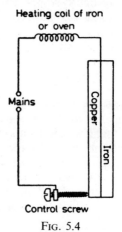

Fig. 5.4

The thermostatic control of a gas oven is based on the principle that on heating, different metals expand varying amounts. In Fig. 5.5, the rod A is made of metal which expands very little on heating, but the outer casing, B, to which it is fixed at one end, expands considerably more, so that it draws the end C of rod A back to cut off the gas supply when the oven reaches the required

HEAT AND TEMPERATURE 33

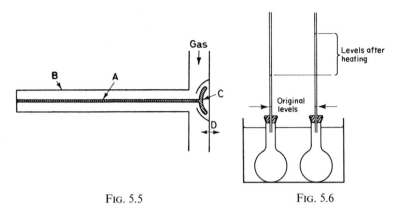

Fig. 5.5   Fig. 5.6

temperature. However, some provision must be made to ensure that the gas flame is not extinguished completely. This may take the form of a by-pass tube of narrow bore, or by suitably shaping the valve C, so that no matter what its position, some gas always can escape past it. The movement of D in the direction of the arrows determines the operating temperature. Seals ensure that gas can reach the appliance only round C or the by-pass.

All liquids do not expand to the same extent, either. If two identical flasks, with protruding tubes, are filled with different liquids (as in the experiment to demonstrate the expansion of a liquid) and placed in a well-stirred heated water bath, it will be seen that the liquids rise to different extents in the tubes (Fig.5.6).

Equal volumes of different gases, on the other hand, expand to the same extent for equal rises in temperature, if the pressure on them is kept constant.

Revision 5
1. State four ways in which heat may affect a body.
2. Write out carefully the principle of the thermostatic control of an electric oven, and describe how it is used.
3. Why is it important that the flame in a gas oven is not extinguished by the regulator?
4. In the experiment to demonstrate the expansion of different liquids (see Fig. 5.6), why should the flasks be identical?

Chapter 6

Measurement of Temperature

Before getting into a bath of water, one usually plunges a finger into the water to test the water's temperature. While the sense of touch—the most primitive method of estimating temperature—is sufficient to ensure a comfortable bath, it would be unwise to use this method to find if an oven is hot enough for baking bread. An oven's temperature is measured by an instrument called a thermometer.

To bake a certain type of bread, an oven's temperature must be approximately 260 degrees. For a number of ovens, each with a thermometer which registers 260 degrees, the baker expects all the ovens to be at the same temperature. This is because thermometers are standardised. This is done by the manufacturer of the thermometer, using a scale of temperature with two fixed points, one of which is the temperature of boiling water, and the other the temperature of melting ice. The water should boil and the ice melt at normal atmospheric pressure.

If the baker were using an old oven he might find that oven thermometers would read differently. Suppose, for example, that the oven was being used to bake the same sort of bread described above, then the thermometer would register 500 degrees. This does not mean that modern bakers use cooler ovens than they used to. On the contrary, they use ovens at the same temperature. The difference in the reading of the thermometers is accounted for by the use of different scales. While the modern baker is used to reading temperatures on the Celsius scale, formerly called the Centigrade scale, the older one has been brought up to read temperatures from a Fahrenheit scale. Both scales use the melting point of ice and the boiling point of water as the fixed points.

In the case of the Celsius scale, the temperature at which ice melts is called 0 degrees Celsius (written 0°C) and the temperature at which water boils is called 100 degrees Celsius (written 100°C). Thus, between the two fixed points on a Celsius scale there are one hundred Celsius degrees. There is another scale, the Kelvin scale, which also has 100 degrees between the melting point of ice and the

MEASUREMENT OF TEMPERATURE 35

boiling point of water. However, the zero on this scale is $-273°C$ ($= 0$ K).

On the Fahrenheit scale of temperature, the melting point of ice is said to be 32 degrees Fahrenheit ($32°F$) and the boiling point of water 212 degrees Fahrenheit ($212°F$) giving 180 Fahrenheit degrees between the fixed points on the Fahrenheit scale.

Therefore, a rise in temperature of 100 Celsius degrees is equivalent to a rise of 180 degrees on the Fahrenheit scale. Thus, 50 Celsius degrees are equivalent to 90 Fahrenheit degrees and 5 Celsius degrees are equivalent to 9 Fahrenheit degrees.

Example:
A rise in temperature of $20°$ is noted on a Celsius scale thermometer. What would the rise be on a Fahrenheit scale?

Since 5 Celsius degrees are equivalent to 9 Fahrenheit degrees then 20 Celsius degrees are equivalent to

$$\frac{9}{5} \times 20 \text{ degrees}$$
$$= 36 \text{ Fahrenheit degrees}$$

Thus, in order to convert a Celsius temperature rise (or fall) to a Fahrenheit temperature rise (or fall) it is necessary to multiply by 9/5. To convert a Fahrenheit rise (or fall) to a Celsius rise (or fall) we must multiply by 5/9.

However, we are not concerned so much with rises or falls in temperatures as with actual temperatures. If, in the above example, both thermometers had been in melting ice at the start, then the Celsius thermometer would have read $0°C$ and the Fahrenheit thermometer $32°F$. A rise of $20°$ on the Celsius thermometer is equivalent to a rise of $36°$ on the Fahrenheit thermometer, on which the recorded temperature would have been $32 + 36 = 68°F$.

In order to convert a temperature from degrees Celsius to degrees Fahrenheit multiply by 9/5 and then add 32.

The student can now see that an old oven registering $500°F$ is at the same temperature as a modern one registering $260°C$.

The next step is to consider how thermometers are constructed to record temperatures.

The simplest of all thermometers is the 'mercury in glass' thermometer which uses the fact that mercury, in common with other liquids, expands on heating.

A glass bulb is blown on the end of a piece of narrow-bore capillary tubing (Fig. 6.1). A funnel sealed to the top of the tube is used to contain the mercury initially. The bulb is then heated to drive out

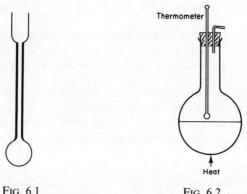

Fig. 6.1   Fig. 6.2

some of the air which expands and, on cooling the bulb, mercury is drawn into the bulb, as water was drawn into the inverted flask in Fig. 5.2 in Chapter 5. Alternate heating and cooling is employed to fill the bulb and tube. The size of the bulb and the bore of the tube are chosen so that if the bulb and tube are full at the highest temperature the thermometer will be required to register, then the mercury level in the tube will be near the bottom at the lowest temperature required. The funnel is then removed with the stem full, and the tube sealed. It is now ready to be graduated.

The fixed points are found by noting the mercury levels when the bulb is placed in melting ice, and in the steam from boiling water which is enclosed in a vessel with only a small outlet (Fig. 6.2). In both cases the level is not marked until it is steady. Steam is used rather than the boiling water itself, as impurities in the boiling water may raise its temperature, but the steam does not contain these impurities and is, therefore, not subject to such temperature variations. Once the fixed points have been marked, the space between them can be divided into 100 divisions, if the thermometer is to read degrees Celsius. Once the correct spacing has been calculated, the thermometer may also be graduated above and below the fixed points.

Mercury is used because it is a good conductor, i.e. heat travels through it easily, so that the whole bulb is quickly at the temperature to be measured. It does not wet glass, leaving a trail behind it. It expands well and has a fairly high boiling point (360°C) and a low freezing point ($-40°$C).

As baking temperatures rarely exceed 360°C, the mercury in glass thermometer is suitable, whereas one filled with, say, alcohol (used in ordinary room thermometers) would be useless, as alcohol

MEASUREMENT OF TEMPERATURE

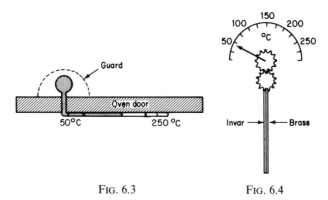

Fig. 6.3   Fig. 6.4

boils at approximately 78°C but is suitable for low temperatures, as alcohol has a very low freezing point.

It would not be practical if the oven door had to be opened each time the thermometer is read. Thus the stem of an oven thermometer has to be bent so that the bulb can be inside the oven and the required portion of the scale outside. The bulb is usually protected with a metal guard which has holes in it to allow the hot oven gas to circulate inside it (Fig. 6.3).

Another kind of thermometer suitable for higher temperatures uses the principle of the compound bar, the amount of bending of which is dependent upon temperature. Bars of brass and invar (an alloy containing nickel and iron) riveted together are very sensitive, and the saw edge at the top of such a bar can be used to turn a toothed wheel with a pointer attached (see Fig. 6.4). Alternatively, the bar can be attached to a lever system which operates a spring.

Another type of thermometer uses the principle of the 'thermocouple'. To demonstrate this, one end of each of two copper wires is joined to a very sensitive instrument which detects electric current. The two free ends are joined to a constantan wire, and the junctions put, one in ice and the other in hot water (Fig. 6.5). A current will be seen to flow. If the temperature difference between the two junctions is raised, then the current will increase. Thus if one junction is kept cool, and the other put in an oven which is being heated, the current will gradually increase. The current will become steady when the temperature of the oven is steady also. By using a mercury in glass thermometer as a comparison it is possible to make a scale on the face of the current measuring instrument, so that it reads temperatures directly. In practice, several junctions are used (Fig. 6.6) to produce a larger current. The current measuring instrument is then made less sensitive and, incidentally, costs less to make.

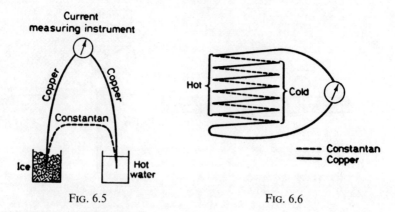

Fig. 6.5   Fig. 6.6

It is often necessary to know the maximum or minimum temperature reached during any given period. Taking the temperature of the human body is an example. The thermometer cannot be read while it is still in the patient's mouth so a clinical thermometer is made with a constriction which prevents the mercury from flowing back into the bulb once the maximum temperature inside the patient's mouth is reached.

A room thermometer to record both the maximum and minimum temperatures during any period is designed as shown in Fig. 6.7.

This is an alcohol thermometer, the purpose of the mercury being merely to push up two springs, B and C, which will move up, but not down.

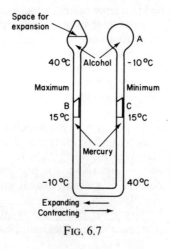

Fig. 6.7

MEASUREMENT OF TEMPERATURE

When the alcohol in A expands, the mercury is pushed further round the bend and B is pushed up. When the alcohol contracts, on cooling, the mercury is drawn up the tube and C with it, but when the mercury moves down again, C is left to record the minimum temperature. The springs may be reset with a magnet.

Revision 6
1. What are the fixed points on the Celsius scale of temperature?
2. Why is mercury a suitable liquid for an oven thermometer?
3. The lowest temperature on a certain oven thermometer is 90°C. When the oven is heated to this temperature will any water in the oven be made to boil? Could an alcohol thermometer be used at this temperature? Give reasons for your answers.

Chapter 7

Heat and Temperature Relationship and Specific Heat

Assume an electric immersion heater is placed in a kettle of cold water and it is switched on. The temperature of the water is recorded after five minutes. The heater is then removed and placed in a large bath containing much more water than the kettle, but initially at the same temperature. Again the temperature is read after five minutes. It is found that the rise in temperature of the water is much greater in the kettle than in the bath.

In a further experiment both the kettle and the bath of water could be boiled by the immersion heater. It takes very much longer to boil the bath of water, i.e. very much more heat has to be given to the bath water in order to obtain the same rise in temperature.

From these two experiments it is clear that the amount of heat given to a body and its temperature rise are not the same. Had two similar kettles been used instead of one kettle and one bath, then the two would have boiled when the same amount of heat had been given to them because the two kettles hold the same mass of water.

We can now conclude that the temperature rise of the water in the first experiment was dependent on the mass of water being heated, or that, for a given temperature rise, the heat needed is dependent on the mass. Double the mass would require double the amount of heat; treble the mass—treble the amount of heat. For a given temperature rise, the amount of heat needed is proportional to the mass heated.

The unit we use to measure heat is the joule, which we met in Chapter 1. Let us consider a temperature rise of $1°C$, and this rise being given to 1 kg of water. The water needs 4200 joules of energy. This could be supplied electrically, mechanically or by burning gas.

The unit of heat used to be the calorie which was the heat needed to raise the temperature of 1 g of water by $1°C$. This was 1/1000th of an old Calorie which was the unit in which the *calorific values* of foods used to be measured. If 10 kg of water are heated through $1°C$, 42 000 J will be needed as each kg of the ten will need 4200 J.

HEAT AND TEMPERATURE RELATIONSHIP

If the 10 kg of water are to be heated through a further 1°C then a further 42 000 J would be required, giving a 2°C total temperature rise for 84 000 J of heat.

From these examples we can deduce that to obtain the number of joules we can apply the expression:

Number of joules = mass of water in kg × temperature rise of water in degrees Celsius × 4200

The examples so far in this chapter have applied to water only. If 1 kg of water is heated by an immersion heater so that the temperature rise is, say, 2° and the same immersion heater is embedded in a block of copper, also weighing 1 kg, until the temperature of that has also risen 2° then the time taken for the copper block to reach the required temperature will be found to be considerably less—in fact only about a tenth of the time taken for the water. Although in this case we have equal masses of water and copper, and the same rise in temperature in each, the copper needed only about a tenth the amount of heat. Now 1 kg of water heated through 1°C needs 4200 J of heat by definition, so 1 kg of copper heated through 1°C needs only 420 J. This figure is called the *specific heat capacity* of the substance. The specific heat capacity of a substance can, therefore, be defined as 'the number of joules required to heat 1 kg of the substance through 1°C.' Its units are joules per kilogram per degree Celsius (J/kg°C).

Example:
How many joules are required to heat 2 kg of lead through 10°C if the specific heat capacity of lead is 125 J/kg°C?

1 kg of lead heated through 1°C needs 125 J
Therefore, 2 kg of lead heated through 1°C needs 250 J
Therefore, 2 kg of lead heated through 10°C needs 2500 J

To get the amount of heat, the mass is multiplied by the specific heat capacity and by the temperature rise.

A simple way of remembering this is:

Amount of heat = MST—which MuST be remembered!

In baking, it is preferable to carry out some processes at certain definite temperatures. For example, dough might be required at 27°C. This dough is obtained by mixing flour and water. As the flour is often at a temperature less than 27°C, warm water is mixed with it to obtain the required temperature. The temperature must depend on the proportions of flour and water which are mixed and also on the specific heat capacity of the flour.

The specific heat capacity of flour is about 1800 J/kg°C, but the student should verify this experimentally—the method used is called the 'Method of Mixtures'. It can be applied in the determination of the specific heat capacity of most substances and the principle used is as follows.

The substance under investigation is mixed with another substance of known specific heat—water if possible, as that has a known specific heat capacity. One of the substances of the mixture is originally cold and the other hot. When they are mixed the hot substance gives heat to the cold one until the whole mixture is at the same temperature. Then provided no heat is lost to the air, the heat lost by the hot substance must equal the heat gained by the cold substance.

In finding the specific heat capacity of a metal, it is usual to heat the metal and lower it into cold water, the volume of which has been found by using a measuring cylinder before transferring to a small can called a calorimeter. The results of such an experiment were:

mass of lead used	= 0.45 kg
original temperature of lead before adding to cold water	= 100°C (heated in boiling water)
volume of cold water	= 200 cm^3
therefore, mass of cold water	= 0.2 kg
temperature of cold water	= 15°C
temperature of mixture	= 20°C
therefore temperature rise of the cold water	= 5°C
therefore temperature drop of the lead	= 80°C
heat lost by the lead	= 0.45 × 80 × specific heat capacity of lead
	= 36 × specific heat capacity of lead
heat gained by the cold water	= 0.2 × 5 × 4200 J
	= 4200 J

Then since the heat gained by the cold water is equal to the heat lost by the lead

$$36 \times \text{specific heat capacity of lead} = 4200$$

$$\text{or the specific heat capacity of lead} = \frac{4200}{36}$$

$$= 117.5 \text{ J/kg°C}$$

HEAT AND TEMPERATURE RELATIONSHIP

To find the specific heat capacity of flour, about 200 cm³ of water are poured into a measuring cylinder or burette and the exact volume noted. Every drop of the water is transferred into a beaker on a gauze on a tripod over a bunsen burner and the water gently heated. It should not be heated too much, as it evaporates more readily at high temperatures and the volume would thus be reduced. A small calorimeter is now weighed and weighed again about half full of flour. A thermometer is placed in the flour and another in the warm water and the temperatures on each recorded. The calorimeter is placed on a cork mat inside another can to avoid heat losses then the warm water is quickly transferred to the calorimeter. The mixture is stirred and the maximum temperature of the mixture recorded.

Specimen results for such an experiment were as follows:

mass of calorimeter = 0.04 kg
mass of calorimeter + flour = 0.25 kg
therefore, mass of flour alone = 0.21 kg
volume of water used = 200 cm³
therefore, mass of warm water
  (density = 1 g/cm³) = 0.20 kg
temperature of cold flour = 14°C
temperature of warm water = 40°C
temperature of mixture = 32°C
therefore, temperature rise of the flour = 18°C
therefore, temperature drop of the water = 8°C
heat gained by flour = 0.21 × specific heat capacity of flour
  × 18 J
heat lost by the water = 0.2 × 4200 × 8

Since the heat gained by the flour equals the heat lost by the water
3.78 × specific heat capacity of flour = 6720

Therefore, specific heat capacity of flour = $\frac{6720}{3.78}$ = 1780 J/kg°C very nearly.

In the above experiments it has been assumed that no heat has been absorbed by the calorimeter. As the specific heat of copper is fairly small, it takes in very little heat, but when an accurate result is required it is assumed that the copper calorimeter has the same temperature rise as the cold body. The heat taken in by the copper (mass × specific heat capacity × temperature rise) is then added to the heat taken in by the cold body, the total equalling the heat given out by the hot substance.

Values of specific heat capacities of almost every known substance have been determined accurately and can be found in a book of constants. The main problem facing the baker, therefore, is not how to determine the specific heat capacity of flour, but how, knowing the specific heat capacity of flour, he can find the correct temperature for the water he is adding to his flour, so that he gets the required final dough temperature.

Flour and water are usually mixed in the proportions 2:1, so in the following example it is assumed that 50 kg (50 litres) of water are to be mixed with 100 kg of flour at room temperature (say 20°C) to obtain dough at 15.6°C. To find the temperature of the water needed we use the following calculation.

The heat lost by the flour = $100 \times 1800 \times 4.4$ J (MST)
$$= 792\,000 \text{ J}$$
Therefore, heat gained by the water must be 792 000 J

Therefore, each kg of water must gain $\dfrac{792\,000}{50}$ J = 15 840 J

But each kilogram of water gains 4200 J when it rises 1°C

Therefore, temperature rise of the water must be

$$\frac{15\,840}{4200}\,°C = 3.77°C \text{ (say } 3.8°C)$$

Therefore the initial temperature of the water must have been
$$15.6 - 3.8 = 11.8°C$$

Because this is lower than the usual summer temperature for tap water and the flour might be stored at an even higher temperature than 20°C, ice is often added to the water to cool it in preparation for making dough.

Had the specific heat capacity of flour been taken as 2100 J/kg°C (half that of water), the heat lost by the flour (and therefore gained by the water) would have been.

$$100 \times 2100 \times 4.4 \text{ J} = 924\,000 \text{ J}$$

Therefore, temperature rise of the water would thus have been

$$\frac{924\,000}{50 \times 4200} = 4.4°C$$

i.e. the rise in temperature of the water equals the required drop in temperature of the flour. This is usually accurate enough for the baker.

The original temperature of the flour − required temperature of mixture
= temperature of mixture − temperature of added water
Therefore, temperature of added water
= (2 × temperature of mixture) − original temperature of flour

If the mixture has to be at 15.6°C for the Chorleywood bread process of mechanical dough development then the temperature of the water needs to be 31.2°C less the temperature of the flour. The temperature of the mixture does not need to be very high, because the work done on the dough by the machine (11 watt-hours per kg) causes a rise in temperature of about 14.5°C so in the worked example the final temperature after working is about 30°C. This is similar to the temperature achieved in conventional doughmaking when the loaves are left to rise in a warm place. Then *warm* water is added to obtain a dough between 27°C and 29°C, the temperature of the warm water being calculated as

(2 × required dough temperature) − flour temperature.

10% is usually added to allow for heat losses.

BOMB CALORIMETER FOR DETERMINING THE CALORIFIC VALUES OF FUELS

We often need to know the number of joules given out by certain foods when they are burnt. For this a bomb calorimeter is used.

A bomb calorimeter consists of a steel vessel to withstand high pressures (Fig. 7.1) with two valve inlets. The food is burnt in C, which is spoon-shaped. Oxygen is pumped in to a pressure of, say, 20 atmospheres and the whole calorimeter is immersed in water which is kept stirred. The food is ignited by passing a current through the coil W, attached to the suspending wires. The rise in temperature of the water is read to 1/100th°C by a very accurate thermometer. Allowance is made for the fact that the calorimeter itself absorbs some of the heat.

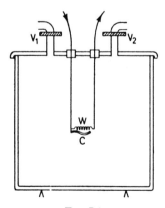

FIG. 7.1

Revision 7
1. Define the term specific heat capacity.
2. How much heat is taken in by the following?
    (a) 20 g of water heated from 25°C to 75°C;
    (b) 50 kg of water heated from freezing point to boiling point;
    (c) 50 g of copper heated from 0°C to 100°C (specific heat capacity of copper = 420 J/kg°C)
    (d) 20 kg of flour heated from 15°C to 25°C (specific heat capacity of flour = 1800 J/kg°C.
3. What is the rise in temperature when 420 J are given to 20 g of water?
4. 2100 J caused a temperature rise of 4°C in water. What was the mass of water used?
5. 100 kg of flour at 15°C are mixed with 50 kg of milk at 39°C. If the temperature of the mixture is 27°C, what is the specific heat capacity of milk, if the specific heat capacity of flour is 1800 J/kg°C?
6. Determine the temperature of the water to be mixed with the flour to obtain a dough temperature of 15.4°C if the flour is at 25°C.

Chapter 8

Changes of State

A Celsius thermometer placed in a saucepan of boiling water registers 100°C. It will continue to register 100°C no matter how long the thermometer is kept in the boiling water. Yet heat must be given to the pan all the time in order to keep it boiling. What happens to this heat?

In Chapter 5, one of the properties of heat was said to be its ability to change the state of a substance. Boiling is a process by which a liquid is changed into a vapour and so in the case of the pan of boiling water the heat is used to change the liquid water into steam which is a vapour. Furthermore, the process stops when the heat is no longer applied, so that heat is necessary to change the liquid into a vapour. Once the source of heat is removed, boiling stops and the temperature of the water slowly drops as some of the heat is given to the surrounding atmosphere. The heat applied to the pan of boiling water must, therefore, counteract heat losses to the atmosphere as well as changing the water into steam.

Why does the heat enable the water to change to steam? In Chapter 1 the three states of matter—solid, liquid and gas—were discussed. In a solid, the molecules of which it is composed are closely packed and their vibrations are strictly within confined limits. In a liquid, the molecules are less closely packed and they can vibrate much more freely (see also Chapter 4) with the result that a liquid will take the shape of the vessel into which it is poured. Being more free the molecules can move more quickly. In a gas or vapour, the molecules are still more free and can move very rapidly and escape right away from the others. The difference between a vapour and a gas is that under pressure a vapour can be liquefied while a gas cannot without cooling it. A solid can be likened to a football crowd packed into a stand with all exits closed. Movement is very limited and no one can move far from his position. The liquid can be likened to the same crowd walking to the station afterwards. They all keep more or less together as they are going in the same direction but sufficient spreading out has occurred for them to move more quickly than they were able to in the stand and to enable them to change positions relative to each other. A few who live nearby may even escape from the crowd altogether into the freedom of their back gardens and parks where they can move as quickly as they like and can get away from all the others. Those with this freedom can then be likened to the molecules of a vapour or a gas.

This 'escaping' from the crowd is exactly what the molecules of a liquid are doing when they become a vapour; exactly what happens when the liquid boils. But to escape they need energy, just as we need food to provide us with the energy to run quicker. This energy is given to the pan of boiling water in the form of heat and without this heat the molecules do not in general have the energy to escape.

Heat energy is needed in the same way when a solid changes to a liquid, but when the reverse process occurs, i.e. a liquid changes back into a solid, heat is given out from the body to its surroundings.

The heat needed to change the state of a body or which is given out when a body changes state is called *latent heat*.

The following is worth remembering:

$$\text{Heat given out} \quad \begin{array}{c} \uparrow \text{SOLID} \\ \text{LIQUID} \\ \text{VAPOUR} \downarrow \end{array} \quad \text{Heat taken in.}$$

However, it should be noted that some substances can change from a solid to a vapour without passing through the liquid state. This is called sublimation.

It can be shown that heat is given out when a liquid changes to a solid by melting some fat or some naphthalene (the main constituent of moth balls) in a test tube then allowing it to cool with a thermometer in it. At first the temperature will drop steadily, then it will remain steady for some minutes. During this time the liquid can be seen changing to a solid, showing that when a substance changes its state it does so at the same temperature. This is the reason for the term '*latent*' *heat* which is usually defined as 'the heat given out or taken in by a body when it changes from one state to another at the same temperature'. However, the fact that the temperature of the body remains constant for a time and yet all the while loses heat to the surroundings (it feels warm to the touch), shows that it must be giving out sufficient heat to make up for this loss, otherwise its temperature would continue to fall.

If a thermometer is placed in a beaker of ice and a bunsen burner is used to heat the beaker slowly a graph of the temperature shown by the thermometer against the time taken would be of the form shown in Fig. 8.1.

CHANGES OF STATE

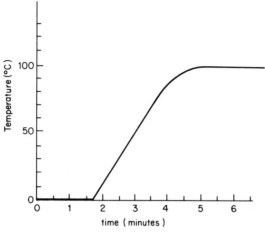

FIG. 8.1

Note particularly that the temperature would remain at 100°C until all the water had boiled away.

Before leaving the subject of latent heat, the student might like some idea of the numerical values of latent heats. To melt 1 kg of ice at 0°C to water at 0°C, approximately 336 000 J are needed and to change 1 kg of water at 100°C to steam at 100°C about 2 million joules are required.

The method of mixtures can be used to find the specific latent heat of vaporisation of water by passing steam into cold water but the steam often contains droplets already condensed to water which make the experiment inaccurate. As an alternative, a simple electrical method which reduces this source of error will be suggested here. An electric kettle or a small immersion heater can be used to heat a known weight of water. It is better to lag the kettle or beaker in which the immersion heater is placed in order to reduce heat losses to the surrounding air. The 'power rating' of the heater or kettle must be known. It is often stated on it and is given in watts. The watt (see Chapter 1) is a unit of power, that is the energy supplied to it per second.

A weighed quantity of water is put into the kettle or beaker and brought just to the boil. As soon as it boils a stop clock is started and the water is allowed to boil away for a timed interval of about 5 minutes. Then the remaining water is weighed again to find the weight of water lost. The following example illustrates an experiment of this sort.

A 2000 watt kettle was kept at the boil for exactly 5 minutes.
The energy given to the water was $2000 \times 5 \times 60$ joules
$$= 600\,000\,J \text{ (or } 6 \times 10^5\,J)$$
The water boiled away in this time $= 0.3\,kg$
Then if 0.3 kg of water needs $\quad 6 \times 10^5\,J$ to vaporise

$$1.0\,kg \text{ of water needs} \quad \frac{6 \times 10^5}{0.3}\,J$$
$$= 2 \times 10^6\,J$$

The specific latent heat of vaporisation of water is thus 2×10^6 joules per kilogram.

The specific latent heat of ice can be found by the method of mixtures by adding small pieces of ice (previously dried on blotting paper) to warm water in a calorimeter. When the ice has melted, the water is stirred and the minimum temperature is recorded. Specimen results are as follows:

mass of calorimeter empty	$= 60\,g$	
mass of calorimeter and warm water	$= 180\,g$	
therefore, mass of warm water alone	$= 120\,g$	$= 0.12\,kg$
mass of calorimeter + water at end of experiment	$= 144\,g$	$= 0.144\,kg$
therefore, mass of ice added	$= 24\,g$	$= 0.024\,kg$
temperature of warm water	$= 28°C$	
temperature of mixture	$= 10°C$	

(This is chosen to be approximately as far below room temperature as the warm water was originally above.)

temperature of ice	$= 0°C$ (assumed)
therefore, temperature drop of water	$= 18°C$
therefore, temperature rise of melted ice	$= 10°C$
therefore, heat lost by the warm water	$= 0.12 \times 18 \times 4200\,J$
	$= 9072\,J$
heat gained by melted ice when heated through $10°C = 0.024 \times 10 \times 4200$	$= 1008\,J$
therefore heat gained by ice in melting $= 9072 - 1008$	$= 8064\,J$

This is gained by 0.024 kg of ice

Therefore, heat gained per kg in melting $= \dfrac{8064}{0.024} = 336\,000\,J/kg°C$

This is the specific latent heat of ice.

This method, with adaptations, could be used to find the specific latent heat of fusion of some fats. In this case the pieces of fat are added to very hot water and the mixture well stirred before taking the final temperature when all the fat has melted.

CHANGES OF STATE 51

When using mechanical dough development the work done by the mixer gives heat to the dough, so causing its temperature to rise by about 15°C. If the ambient temperature is too high it may be necessary to cool the water, by adding ice, before dough making, to achieve a correct final temperature.

Latent heat of evaporation forms the basis of cooling by mechanical refrigeration (widely used in the production, as well as the preservation, of foodstuffs). As previously explained, the addition of latent heat to a liquid, at its normal boiling point, will bring about a change in state and vapour will be formed. The temperature at which this change occurs depends upon the prevailing pressure. For example, water boils at 100°C at atmospheric pressure, but by raising its pressure to 1.5 times atmospheric pressure the corresponding boiling point is increased to 111°C. Conversely if the pressure is reduced (using a vacuum pump) to one hundredth of atmospheric pressure water will boil at 7°C and can be used for carrying out effective cooling.

Fluids more suitable than water are used in a modern refrigeration plant. They are all inert, at normal temperatures, and are neither toxic, inflammable, corrosive nor irritating. These fluids, termed refrigerants, are identified by a number which enables the chemical structure to be determined. *Refrigerant 12* (dichlorodifluoromethane) is commonly used in domestic refrigeration.

Synthetic refrigerants have replaced the organic fluids widely used in the past, except that ammonia is still in common use for large cold stores. They are obtainable from refrigeration equipment suppliers at low cost, in disposable containers holding either 0.5 or 1.0 kg. With the aid of a small container of *Refrigerant 12* and a few other items that are normally available in the physics laboratory, the principle of the modern refrigerator can be effectively demonstrated.

A piercing valve is attached to the cylinder outlet which is then inverted and supported as shown in Fig. 8.2. A small bore copper tube is used to carry the liquid refrigerant into a coil placed in a glass beaker partly filled with water.

The piercing valve is opened to allow the liquid to flow into the coil where it evaporates and extracts its latent heat from the water. Vapour flows from the open end of the tube and the heat required is taken from the water, which rapidly freezes on the outside of the immersed coil. This vapour is completely safe provided that it does not come into contact with a naked flame or electric heating element. (At high temperatures decomposition of the refrigerant vapour occurs and the products have a pungent odour which is harmful in high concentrations.) The hazard is, however, no greater than

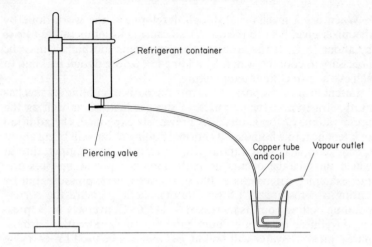

Fig. 8.2

that presented by the everyday aerosol spray, in which the propellant used is invariably *Refrigerant 12*.

In a complete refrigeration system the vapour is extracted from the coils by a compressor and delivered into a heat exchanger where it is cooled, by air or water, so that liquefaction occurs, to complete the cycle shown in Fig. 8.3.

Liquid refrigerant flows into the evaporator (see Fig. 8.3), controlled by the expansion valve, which also maintains the essential pressure difference between the low pressure (evaporating) and

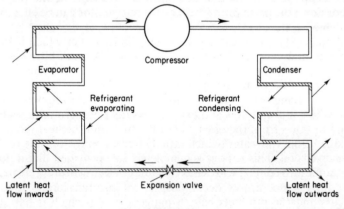

Fig. 8.3

high pressure (condensing) sides of the system. Evaporation occurs at low temperature and pressure in the evaporator coil, which is located in the space or liquid which it is required to cool.

The compressor, normally driven by an electric motor, draws off this vapour from the evaporator and delivers it into the condenser at high pressure and thus a correspondingly high boiling point.

The condenser is then arranged to cool the high pressure vapour below its boiling point, so that condensation occurs and a further supply of liquid refrigerant is available to feed the expansion valve.

Refrigeration plant is used in bakeries to store perishable raw products, prevent wastage during interrupted processing and to store the finished product.

It has already been mentioned that the average energy of the molecules in a liquid increases with rising temperature and thus with a higher temperature there are more molecules with sufficient energy at the surface to escape from the others. Thus, the higher the temperature the faster the evaporation.

However, unless there is a draught or wind to blow away the escaped molecules, they tend to remain near the surface of the liquid from which they have escaped. There is a limit to the amount of water vapour that the atmosphere can hold and when that limit has been reached we say the air is saturated and it cannot hold any more escaped molecules. If more molecules try to escape, then others must return to the liquid state to make room for them.

The amount of vapour that the air can hold increases with rising temperature. Thus, as a vessel of liquid is heated, so more molecules escape. The escaped vapour has a pressure called vapour pressure, but atmospheric pressure, being greater, keeps the vapour close to the surface of the liquid. However, as the temperature rises still further, the vapour pressure increases until it equals atmospheric pressure. The vapour can then escape and more can take its place. At this point boiling is said to occur and the escape of molecules takes place from the whole body of the liquid. A liquid boils when the pressure of its vapour equals atmospheric pressure.

It can now be seen why the boiling point of a liquid depends on the pressure of its surroundings. In a pressure cooker, the thick enclosed pan withstands the high pressure caused by the evaporating, and later boiling, water. As the pressure builds up, so the temperature to which the water must be raised before it boils increases. Eventually the pressure is such that the escape valve opens sufficiently to maintain the vessel at the required pressure, which is usually about 100 000 newtons per square metre above atmospheric pressure but in some pressure cookers this can be varied. A pressure of 30 000 newtons per square metre is sufficient for bottling fruit, for example. The food is cooked in water considerably above the normal boiling temperature due to this increased pressure against which the escaped molecules have to push equally before boiling occurs.

We have now seen that evaporation occurs more easily when the temperature is higher and also when there is a wind or draught. This is why a windy, sunny day is best for drying clothes. Also, if the clothes are spread out to expose the greatest surface area, more water vapour can escape. However, there is another very important factor which affects the speed with which clothes dry, and that is the amount of water vapour already in the air. If the air is already saturated with water vapour then it cannot take any more from the clothes. If it holds very little then it will readily take in more.

The quantity of moisture that air can hold in the form of water vapour depends upon its temperature. Warm air has a high capacity for absorbing water vapour but this is progressively reduced as the temperature is lowered. Condensation-type problems occur in buildings when warm air, with a high moisture carrying capacity, comes into contact with a cold surface which is below its *dew point*. (The dew point of air may be expressed, in simple terms, as the lowest temperature to which a sample of moist air may be cooled before condensation starts to occur.)

Air conditioning plants are used in food production to control both the temperature and the moisture content of the air. The latter is expressed as the relative humidity or percentage saturation and is the ratio of the mass of moisture present to that required to completely saturate the air.

Bakeries are air conditioned to avoid conditions of high humidity, which gives rise to problems of flour becoming moist and sticking to the milling machinery and low humidity, which causes hardening of the surface of the dough, which gives a bread that has hard lumps or holes present. High humidities are also uncomfortable when working and accelerate corrosion problems in the bakery.

The humidity of an atmosphere may be determined by two thermometers placed side by side, the bulb of one of which is surrounded by wet gauze kept wet by dipping into a small can of water (Fig. 8.4). The rate at which the water evaporates from the gauze depends on the humidity of the atmosphere. Since evaporation causes cooling, the bulb of the thermometer covered by the gauze is cooler than the other one and a lower temperature is recorded. Thus the lower the humidity the greater the temperature difference between the two bulbs of this *hygrometer*. Tables have been prepared from which, with a knowledge of room temperature and the differ-

Fig. 8.4

ence between the readings, the relative humidity may be read directly.

Before reading, the wet and dry bulb hygrometer is swung round the room like a football supporter's rattle to make sure the average humidity of the room is found and not that of one damp corner.

When hair becomes moist its length increases slightly. In the hair hygrometer a hair, fixed at one end, is coiled round an axle of a wheel and attached to a spring which keeps it taut. Any movement of the hair due to a change in the humidity of the air will be reflected in the movement of a pointer turned by the wheel. This is illustrated in Fig. 8.5

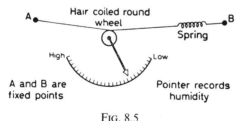

Fig. 8.5

Revision 8

1. Explain what happens when latent heat is (a) supplied to a liquid, (b) removed from a liquid.
2. Crushed ice at a temperature of $-5°C$ is placed in a beaker and then slowly heated to $+5°C$. Plot a simple graph, with temperature as the vertical axis and time as the horizontal axis, to show the changes that occur.

3. When the following changes of state occur explain whether heat has to be added or removed. (a) Solid to liquid. (b) Vapour to liquid. (c) Liquid to solid. (d) Liquid to vapour.
4. Describe how the boiling point of a liquid is affected by changes in pressure.
5. Explain why air humidity is important in a bakery, how it is controlled and how it can be measured.
6. Mechanical refrigeration is an essential feature of complete air conditioning. Explain the functions it has to perform in the treatment of air supplied to a bakery.

Chapter 9

Transmission of Heat

Heat can be transmitted in three ways—by conduction, convection, and radiation.

When a body is heated, as explained in Chapter 5, the energy given to it causes the molecules of which it is composed to vibrate more rapidly. These vibrations are, of course, too small to be seen, even with a powerful microscope. If a hot body is placed in contact with a cold body some of its energy is transferred to the cold body whose molecules in turn begin to vibrate more quickly. This transfer of heat energy by contact from one body to another, and from one part of the same body to another part, is called *conduction*.

Conduction can take place in solids, liquids and gases, but cannot take place in a vacuum where there is no matter present to transfer the energy. It is the only way heat can be transferred in solids. Transference of heat by conduction can be illustrated by a number of examples from everyday life. If one end of a poker is placed in the fire, the other end soon becomes hot; the heat has been transmitted by conduction from one end of the poker to the other. When a hot stove is touched, the heat is rapidly transferred to the hand. But heat passes from the hand when it touches a colder body such as a refrigerator. In stirring the contents of a heated saucepan, a wooden spoon is usually used since the handle of a metal one becomes too hot to hold. Substances which conduct heat easily are called good conductors and those which do not conduct it easily are bad conductors. Most metals are good conductors of heat. That is why metal below body temperature always feels colder than, say, wool at the same temperature. Metal takes heat from the body more rapidly than wool.

Good conductors are used for hotplates, saucepans, shelves and sides of ovens, and for baking trays and tins, so that the heat is easily transferred to the bread or confectionery being baked. A metal spoon is sometimes placed in a glass before pouring in a hot liquid, so that the spoon quickly conducts the heat away from the liquid, thus preventing the glass from cracking. If the spoon were not used, the inside of the glass would be heated and would expand. As glass is such a poor conductor of heat, the outside would not get heated and would therefore not expand, causing a strain to be set up in the glass which would cause it to crack.

Some metals are better conductors of heat than others. This can be demonstrated in the laboratory by an apparatus, designed by Ingen-Hauz, as shown in Fig. 9.1. Several specimens of the metals under investigation in the form of solid rods of identical dimensions protrude through corks into a metal box containing water which is being heated. The rods are coated with wax and lead shot is stuck into the wax all along the undersides of the rods.

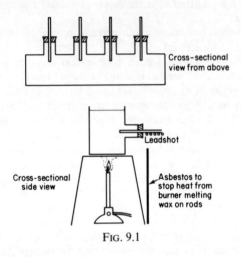

Fig. 9.1

As the heat travels along the rods, the wax melts and the lead shot falls off. When no more lead shot will fall from any of the rods, it will be noted that the wax has melted much further along some rods than others.

The laboratory apparatus does not usually contain a rod of silver, one of the best conductors, but it usually contains iron, brass, copper and aluminium rods. It will be noted that the wax has melted further on the copper than on the other rods, so our grandmothers had good reason to use copper pans and kettles. However, aluminium is nearly as good a conductor and, because of its lightness and cheapness, this metal is now preferred for modern kitchen equipment. Brass will be found to be a better conductor than iron.

Some laboratories have an Ingen-Hauz apparatus which includes rods of glass and wood. On these the wax hardly melts at all, showing that glass and wood are bad conductors of heat.

A further experiment to show the heat-conducting powers of metals can be performed with a piece of fine copper wire gauze and a bunsen burner. If the gauze is held over the flame of the burner, the flame will not pass through the gauze. This is because the gauze conducts the heat and does not allow the gas above it to reach its ignition temperature—the temperature at which it would burn. This principle was used in the last century by Sir Humphry Davy in his safety lamp for coal mines. An oil lamp was surrounded by a cylinder of wire gauze so that dangerous gases in the mines were not allowed to reach ignition temperature. With a little care, a student should find that if he holds a piece of gauze above an unlit burner, and puts a lighted match above the gauze and turns on the gas, the flame will not extend below the gauze, which this time has prevented the gas below it from reaching ignition temperature.

Bad conductors, as well as good conductors, have their uses in our everyday lives. Without our woollen clothes, which are bad conductors of heat, we would feel very cold in winter, as the bad conductor prevents the heat of our bodies from escaping. Similarly, woollen blankets and eiderdowns keep us warm at night. The reason that woollen materials are such poor conductors is that they are hairy or fibrous and trap air in the fibres and hairs of the material. It is the air which is a very poor conductor rather than the wool itself and the air prevents the heat from escaping through the woollen material or through any other material in which air is trapped. Lagging put round hot-water tanks and pipes to prevent loss of heat is not usually made of wool but of some other loosely woven or packed material which traps air.

Asbestos and cork are two other poor conductors. Asbestos sheet and cork dust are used for lagging. Bakery ovens have a layer of a bad conductor between the inner and outer casings to prevent loss of heat. This must be rigid or well packed to prevent movement when the door is banged. Its specific heat capacity must be such that the oven is not cooled too much when cold food is put in, yet the oven must not take too long to heat. Cork mats are used to prevent polished tables from being spoiled by heat. A scrubbed wooden surface in a bakery is better than an enamelled one for putting hot baking tins on as the wood does not quickly take heat from the metal. Wooden tables also prevent much heat being taken from the dough when it is being prepared on the table. Handles of pans and kettles are made of poor conductors.

Nearly all liquids are poor conductors of heat. The fact that water is a poor conductor of heat can be demonstrated by weighing a piece of ice and dropping it to the bottom of a test tube of water. If the top of the test tube is now heated, the water at the top can be made

60 PHYSICS AND CHEMISTRY OF BAKING

to boil, while the ice remains unmelted. This shows that the water has not conducted the heat well.

After this last experiment, the student may be wondering how it is that a kettle of water on the stove becomes evenly heated. The reason for this can be demonstrated using a flask of water and a crystal of colouring material, perhaps potassium permanganate, carefully placed in the bottom of it, by dropping the crystal through a glass tube.

If the flask is heated, it will be noticed that pink streams of liquid rise from the crystal and take the paths shown in Fig. 9.2. The water nearest the flame becomes heated by conduction. As it becomes heated, it expands and, therefore, becomes less dense. It rises above the more dense cold water in the same way as less dense cream rises to the top of milk. As the hot water rises, cold water falls to take its place; this in turn, becomes heated, so a continuous flow of water is set up. The movements of liquids in this manner are called *convection currents*.

Convection has been defined as 'the movement of a body carrying its heat with it'.

Convection can take place in liquids and gases, but not in solids or in a vacuum. Without it, ventilation problems would be much greater, and even the problem of providing hot water for a house would be difficult.

In the ordinary domestic hot-water system, water in the boiler, A in Fig. 9.3, is heated and rises to the top of the hot-water tank B from where it is drawn off for the bathroom and sink. Cold water from the cold-water tank C takes the place of the water drawn off

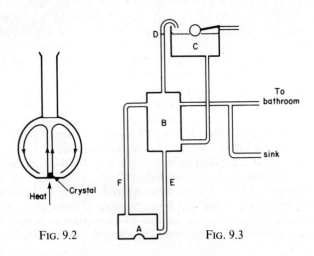

FIG. 9.2 FIG. 9.3

and this runs down tube E to be heated in the boiler, and in turn rises to the top of B. If no water is drawn off, convection currents take the path AFBEA as the water becomes hotter and hotter. D is a safety tube in case of boiling.

A second tank inside B is fitted in centrally heated buildings and a pump circulates hot water from it round the building. This pump is operated by a thermostat.

For times when it is not convenient to keep the boiler alight, an immersion electric heater can be fitted in the bottom of the hot tank, to provide hot water.

This system has the advantage that a large supply of water is readily at hand; but it does involve keeping a boiler alight continuously. Lagging of the pipes and tank can help to prevent heat losses, but for occasional use, instant hot-water heaters of the gas or electric variety prove more economical. In these, only as much water as is immediately required is heated, although the flow from them must, of necessity, be comparatively slow.

It is possible now to explain why it was necessary to heat the test tube at the top, when showing that water was a bad conductor of heat. Convection currents had to be avoided.

Convection currents in gases can be demonstrated by a simple laboratory experiment. A closed cardboard box has two round holes in it, through which protrude two pieces of wide glass tubing, to act as 'chimneys'. A candle is lit under one chimney, B in Fig. 9.4,

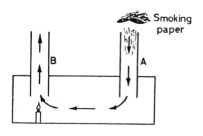

Fig. 9.4

and a piece of smoking paper is held over the other, A. The smoke is seen to be drawn down chimney A and out of chimney B. This is because the candle heats the air around it, which thus rises up B, and air and smoke is drawn down A to take its place.

Exactly the same effect is used in the ventilation of a room. The hot air which rises is let out at the tops of the windows and doors and through the chimney and cold air is drawn in through the

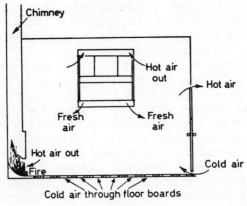

Fig. 9.5

floor boards, under the door and through the lower parts of the windows (see Fig. 9.5).

Ventilation shafts above ovens serve the same purpose as the chimney above the coal fire. Most modern bakeries have electric fans to keep the air circulating, or ventilators in walls which deflect the air upwards as it comes in so that it mixes with the warm air gradually and does not cause a draught either on the operator or on the dough.

Convection currents play a very important part in the heating of ovens. The heated air and flue gases circulate all round the ovens by convection currents and, in order that they should give up as much of their heat as possible before going out into the atmosphere, baffle plates are fitted to help the gases to circulate. These are usually metal plates with holes in them. The actual construction of ovens will be described more fully later when all three methods of transference of heat have been described.

The earth is heated by the sun. As there is a vacuum over the great majority of the distance between the earth and the sun, heat cannot be transferred by either convection or conduction, since both these processes require the presence of molecules. There must therefore be a third way by which heat is transferred. It is called *radiation* and is defined as 'the transference of heat from one body to another, without the intervening space being affected'.

Heat from coal and electric fires reaches us by radiation. If we sit in front of a blazing coal fire, we feel very hot, but if a screen is placed between us and the fire we immediately feel less warm. This shows that we were not receiving heat from the air around us, the temperature of which would be found to be comparatively low, but directly from the fire, by radiation. So-called radiators, on the other hand, do not give out a great deal of radiated heat, but heat the air of the room by convection currents. Some heat from a coal fire reaches us by conduction and convection too, not only from the fire, but from objects in the room which have received radiated heat from the fire. Direct radiation, however, is the main means of transference.

Experiments on the radiating power of different surfaces can be performed using a metal box, known as Leslie's cube (Fig. 9.6), which has its four vertical faces painted or coated respectively dull white, silvered, dull black, and shiny black. Boiling water is poured into the box. A thermopile, as was illustrated in Fig. 6.6, is placed a fixed distance from each face and the deflection on the galvanometer noted. It is found that the deflection is greatest on the thermopile placed in front of the dull black surface and least on the one in front of the silvered surface. This shows that dull black surfaces are the best radiators of heat and shiny ones are the poorest.

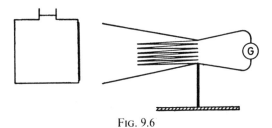

FIG. 9.6

Therefore, a dull black surface is preferable for the inner walls of an oven, as they are required to radiate as much heat as possible. To investigate which kinds of surfaces absorb radiation better, the bulb of one of two identical thermometers is blackened with smoke from a candle, and a bunsen burner placed between them, an equal distance from each bulb. The temperature recorded by the thermometer with the blackened bulb rises more rapidly than the other, showing that the dull black surface is a better absorber of radiation than the shiny surface of the glass of the other thermometer. Thus good emitters of radiation are good absorbers, and bad emitters are bad absorbers.

Bakers wear white overalls because these do not absorb heat as readily as dark ones. Similarly people wear white clothes in summer to absorb less radiated heat, and dark in winter to absorb more. New baking trays are often put in the oven for some time before use, in order that their shininess should wear off; they will then absorb radiated heat better.

Shiny surfaces are poor absorbers of radiation because much of the heat is reflected from them, rather than being absorbed. In the 'radiant' type of electric fire, the reflector should be kept as shiny as possible, so that as much heat as possible is reflected.

The 'thermos' or vacuum flask (Fig. 9.7) makes use of the fact that silvered surfaces are good reflectors and bad radiators and absorbers of heat. A double-walled, glass vessel has the outside of its inner wall silvered to reduce radiation to a minimum, and the inside of its outer wall is also silvered so that as much as possible of any radiation falling on it is reflected, rather than absorbed. The

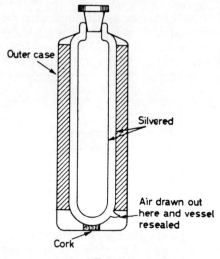

FIG. 9.7

space between the two walls is evacuated of air (i.e., a vacuum is left between the walls) and the vessel is sealed. The absence of any material medium between the walls prevents transfer of heat by conduction and convection currents, and hot liquids will remain hot, or cold remain cold, for up to 24 hours. The vessel is placed in an outer case for protection. A plastic stopper at the top—a bad conductor—keeps the liquid sealed in the flask.

CONSTRUCTION OF OVENS

Having dealt with the ways in which heat can be transferred, it is now possible to consider how each is utilised to its best advantage in the construction of ovens and how fuels are used to obtain heat energy.

There are two main classes of ovens—those which are internally heated, and those which are externally heated. In the internally heated type, the products of combustion, whether from coal, coke, oil or gas pass directly into the oven chamber and circulate by means of convection currents, whereas in the externally heated type they do not enter it, although in some cases they surround chambers in which the bread is baked (Fig. 9.8).

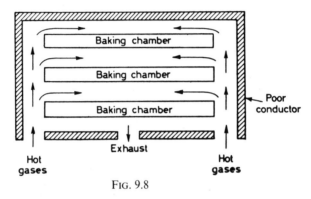

Fig. 9.8

In the side-flue oven, a form of the internally heated type of which there are still some in use in small bakery concerns, the fire is lit in one corner and a flue with a controlling damper is built in the opposite corner at the top. As the heating would not be even with the coal or coke actually burning, these ovens are made of brickwork and clay which will store the heat and, after the fire has been damped down, the stored heat is given to the bread by radiation and also by a certain amount of convection and conduction. These ovens have the disadvantage that they tend to pollute the bakehouse air with dust and fumes and must be reheated before further batches of bread are baked, although they can be used for cakes, which require a lower temperature, afterwards. Also, they are only suitable for a small work load where the bread is removed by wooden or metal spades or peels. In common with other ovens which require the use of peels they are called 'peel ovens'.

Where larger batches of bread have to be baked, the bread may be loaded on drawplates, which can be drawn in and out of the oven on runners, and in the larger baking concerns the travelling belt principle is employed. Alternatively the proved dough can be loaded on to racks and wheeled into the oven. There they may be rotated on a turntable (rotorack ovens) or transported through the oven tunnel (tunnelrack ovens) whilst controlled air circulation helps to ensure even firing of all the work load. The inner and outer shells of the oven chamber are made of stainless steel.

In the reel oven the dough is carried on plates which rotate vertically about a common spindle or reel, not unlike the big wheel at a fun fair, so that each plate can be stopped opposite the oven door. Reel ovens are economical of space but are subject to mechanical faults.

These ovens are of the external heating type with a separate combustion chamber and heat exchanger. Steam-tube heating can be used for all types of ovens. Sealed tubes containing water are used; one end of each tube is heated by the furnace which may be fired by oil, coal, coke or gas, and the other end protrudes into the baking chamber. This method gives uniform heating.

Yet another form of external heating which can be used for peel ovens and for some drawplate ovens is the hot-air system. In this, the products of combustion travel round the baking chambers (Fig. 9.8). They are usually fired by oil or gas.

Convection currents are set up in the direction shown. The walls of the chamber receive heat directly by conduction from the hot air. The dough receives heat from the trays which are good conductors. The walls give out further heat by radiation, which is received by the trays and the upper surface of the dough. Further convection currents are set up inside the actual baking chamber, as the air inside the chamber receives heat from the walls by conduction. Electricity is also used for heating ovens in the baking industry.

Typically, oven temperatures are about 250°C. Baking takes about 30 minutes in the oven. When Vienna goods are being manufactured, steam is fed to the oven to increase the humidity, particularly at the commencement of baking.

Revision 9

1. Write brief definitions of the terms 'conduction', 'convection', 'radiation'.
2. By what processes does heat reach a hand held (a) above, (b) below a hotplate?
3. Why are tea-cosies more often used with dull earthenware teapots than with bright silver ones?
4. Compare the advantages and drawbacks of metal, marble, and wooden working surfaces as used by the confectioner.
5. A new shiny tin and an old black one are both heated to the same temperature. Compare the ways in which they subsequently lose heat; which one becomes cool more quickly?

Chapter 10

Light and Waves

Everyone knows that, without the aid of instruments, it is impossible to see round corners. If a straight line cannot be drawn from an object to one of our eyes without passing through an opaque material, then it is impossible for us to see the object. Thus, light must travel in straight lines. This can be further proved by holding two cards with small holes in them between our eyes and a lamp. If the holes are in line with the lamp and our eye then the rays of light (as we call the paths taken by the light) can pass through and we see the lamp. If, however, they are not in line then the lamp cannot be seen.

What, then, is light? It is something given off by certain bodies, travels in direct lines, and causes us to experience the sensation of vision. We call the 'thing' that is given off 'energy'. It is not really very different from the heat energy which we have discussed. Any body which gives off light energy is vibrating and it sends vibrations out into the air around it. The vibrations are similar to the waves on a pond when a stone is thrown into it. The stone only hits the water in one spot, but the ripples spread themselves out until they hit the edges of the pond. Similarly light waves from a body giving out light spread out evenly until they hit an opaque object.

Up to now, reference has been made to 'bodies which give out light'. Such bodies are called luminous bodies and examples are the sun, a candle, a fire and an electric light bulb. They are, generally in fact, things that are burning, so that they get their energy chemically, or bodies that get their energy electrically.

The moon is not a luminous body. Light from the sun strikes it and bounces off it again thus outlining it, for light from the sun that has not hit the moon passes straight on and does not hit our eyes, whereas rays that have bounced off the moon can be received by the eye.

Most of the objects around us are similar to the moon in this respect, and are called non-luminous bodies. We only see light that has 'bounced' off them. If we put a table in a darkened room, that is, we cut off all light from luminous bodies, we cannot see the table. It is therefore 'non-luminous'. A black non-luminous body absorbs all the light that falls on it and, therefore, we only see a black body outlined by other bodies.

One often hears the expression 'all the colours of the rainbow'. Ordinary light given out by the sun or a flame really consists of light of many different colours. These colours are split up when a rainbow is formed. A similar effect can be obtained by viewing a white card through a triangular glass prism. All the colours of the rainbow—violet, indigo, blue, green, yellow, orange and red can be seen. The vibrations in the luminous body are slightly different for each colour, but as most luminous bodies give out all these vibrations, we receive a combination of all the colours which we call 'white' light. It can be shown that all the colours of the rainbow combine to make white by rapidly rotating a disc on which are painted the seven colours (Fig. 10.1). The disc appears white, for we see each colour for a fraction of a second after it has hit our eye and so all the colours are seen together.

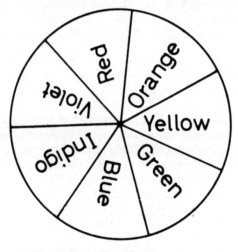

Fig. 10.1

When we say an object is red, what we really mean is that it absorbs every colour of the rainbow except red. Similarly a piece of blue cellophane absorbs all colours except blue, which it allows to pass. A blue book absorbs all coloured rays except blue rays, which it allows to bounce off.

The term 'bounce off' has been used rather loosely up until now.

LIGHT AND WAVES 69

Light strikes the surface of a book from a luminous body and then the book sends it off in all directions. However, if the light strikes a mirror there are definite laws which govern the way it travels after hitting the mirror. A mirror is called a reflecting surface and, for simplicity, we shall consider one ray of light striking it.

When the ray strikes the mirror it bounces off so that the ray that comes off (called the reflected ray) makes exactly the same angle with the mirror as the ray which strikes it (called the *incident ray*).

This is one of the laws of reflection and is obeyed by rays striking mirrors and highly-polished surfaces. The angle between the incident ray and the normal (line at right angles) to the surface is called the *angle of incidence*, and that between the reflected ray and the normal is the *angle of reflection*. These two angles are equal (Fig. 10.2).

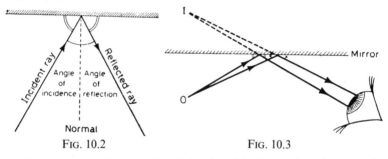

FIG. 10.2   FIG. 10.3

Consider an object O (Fig. 10.3) placed in front of a mirror, and consider two rays from it, each of which obeys the law of reflection, and let these two rays strike our eye. The eye receives two rays in two directions, and the brain imagines that they have come straight from the object and therefore thinks the object is at I. I is called the image of the object and can be defined as the 'point from which the rays appear to come'. This accounts for the fact that a person can see his reflection behind a mirror or below the surface of a pond.

While light 'bounces off' a number of bodies, there are certain substances through which it will pass; these are called transparent substances. However, rays of light do not usually pass straight through such substances unless they strike the surface normally (at right angles to the surface). Imagine pushing a lawn-mower at an angle across a path to a lawn. Let the arrowed lines in Fig. 10.4 represent the tracks of the mower wheels initially at X and Y, which will be at A and B at the same time. Now wheel Y can still proceed on the path, but wheel X has to travel on the more resistant grass. Y can therefore travel a longer distance than X for the same push and will reach C while wheel X only travels to D before they go off

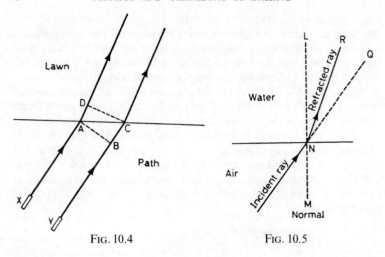

Fig. 10.4   Fig. 10.5

together on the lawn. Thus the mower turns as it goes from one surface to the other.

In the same way light turns or bends as it passes from one material or medium to another. Light travels easily through air, but less easily through glass and water which are called more dense media. To discover which way light bends when it reaches a surface between two media, a normal is drawn where the ray touches the surface (N in Fig. 10.5). L M is then the normal. If the ray travelled straight on, then it would take the path N Q, but if it goes from a less dense to a more dense medium (perhaps from air to water, for example) then it bends towards the normal and takes the path N R.

On passing from a more dense to a less dense medium, rays of light bend away from the normal. As an example, Fig. 10.6 shows

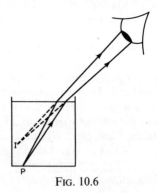

Fig. 10.6

LIGHT AND WAVES

why a swimming bath appears shallower than it really is. Rays from P are bent away from the normal at the surface of the water. An observer imagines that they have come from I which is in line with the rays which strike his eyes. This bending of light when it passes from one medium to another is called *refraction*. If a penny is placed on one side of the bottom of a shallow tank of water and viewed from directly above, then it will appear to be in less depth of water than it actually is. If now you move your head across the tank to a point above the opposite side (see Fig. 10.7) then at some point during the movement the penny will suddenly disappear, not to return unless the head is moved back past this point.

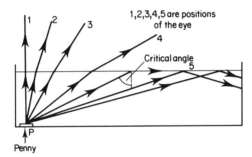

Fig. 10.7

What has happened is that the rays of light given off from the penny, once they strike the surface at a certain angle (called the *critical angle*), are no longer refracted at the surface. Instead they are *reflected* back into the water.

When the eye is in positions 1, 2, 3 or 4 in Fig. 10.7 the rays will be received by it after refraction at the surface. Once it reaches position 5, however, it cannot receive any light, as the refracted ray now runs along the surface. For angles of incidence greater than this critical angle, the rays are reflected.

It should be noted that this reflection, called total internal reflection, occurs only for rays of light travelling in a *more dense* medium, such as water or glass, and striking the surface of a less dense medium, as only in these circumstances is the angle between the refracted ray and the normal greater than the angle between the incident ray and the normal.

If the water in the tank were replaced by different strengths of sugar solutions, it would be found that the critical angle would be different for different strengths of solutions. This provides a means of finding the strengths of sugar solutions which are too sticky to be tested by the hydrometer method.

The instrument used is called a *refractometer*. It will be noticed in Fig. 10.7 that the critical angle is the angle of incidence for which the refracted ray grazes the surface of the water. Since all rays of light have reversible paths, a ray which grazes the surface would enter the water at the critical angle. This principle is employed in the Pulfrich refractometer.

The sugar solution is contained in a small vessel A, standing on a glass block B (see Fig. 10.8). Since light rays which graze the surface of the less dense medium enter the more dense medium at the critical angle, then a beam of light, directed as shown to graze the surface between A and B in medium A, will enter B at the critical angle C. From the angle at which the light emerges at D, the strength of the sugar solution can be calculated.

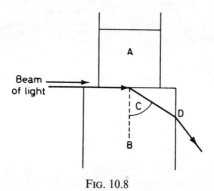

FIG. 10.8

Yet another method of determining the strength of a sugar solution is by using polarised light. The energy given off by luminous bodies is in the form of vibrations. These vibrations have been compared with the ripples on a pond and can also be likened to the vibrations of a skipping rope held between two people, one of whom is moving the hand holding the rope backwards and forwards at various angles to the vertical. The lines in Fig. 10.9 represent the different paths taken by the moving hand.

The vibrations of ordinary light are in all directions, but if some way could be found of cutting out all the vibrations except those represented by A B, then the light would be plane polarised; that is, the vibrations would be in one plane only.

Polarised light may be obtained by reflecting ordinary light at an angle of incidence of 56° or by passing ordinary light through

LIGHT AND WAVES

Fig. 10.9

certain crystals or a substance called Polaroid. When the polarised light energy from a piece of polaroid is viewed through a second piece of polaroid (Fig. 10.10 (i)) it will be found that if the second piece of polaroid is turned, then in one position the light is extinguished and a quarter of a turn later it is at maximum brightness. When the second piece is held in such a position as to give extinction of light and a trough containing sugar solution is placed between the two pieces of polaroid as in Fig. 10.10 (ii) then the second piece (the analyser) will have to be turned through an angle before

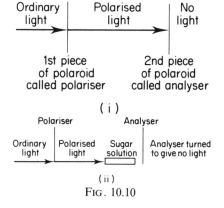

Fig. 10.10

the light is again extinguished. For a given length of trough a table of strengths of sugar solution against angle turned is available.

Some sugars require the analyser to be turned one way and others the opposite way. Sucrose and invert sugar rotate the light in different directions.

Light is only one form of energy that travels in waves. For example, if we consider the waves made by a stone falling into a pond then there are a few measurements we could make on the waves which spread out around it. We could measure the distance between two adjacent crests. This distance is the *wavelength*. We could also find how far the outermost wave travels in 10 s and hence calculate the distance it travels in 1 s. This is said to be the *velocity* of the wave. We could also count how many waves pass a certain point in 10 s and so calculate the number of waves which pass each second. This is said to be the *frequency* of the waves. Repeating this experiment several times, we should find that each time,

$$\text{velocity} = \text{wavelength} \times \text{frequency}$$

A relationship of this kind is true for all forms of wave motion. In some cases the waves are at right angles to the direction of travel, as in the case of ripples on a pond, but there are also waves which move in the same direction as the energy. The blast of a whistle sends sound energy as waves of changing pressure, moving away from the source of the sound. Like light, sound waves can be reflected and refracted. The speed of sound depends on the medium in which it is travelling and its temperature.

There is a system of waves, known as the *electromagnetic spectrum*, which have the property that they all travel with the same velocity (about 300 million metres per second). Light is a part of this spectrum. The different colours of the light which we see are due to the different wavelengths, and so different frequencies too, of the spectrum of visible light. But the spectrum extends beyond visible light: those with shorter wavelengths include gamma rays from nuclear reactions, X-rays and ultra-violet light. The wavelength of violet light is shorter than that of green light which is again shorter than that of red light. Infra-red rays have longer wavelengths than red light. Of longer wavelength still are microwaves and radio waves. In all these cases the velocity of the waves is constant and is equal to wavelength × frequency. Waves with very short wavelength, such as gamma rays and X-rays, are able to penetrate some solid materials, some waves whose lengths are not so short can be detected by our eyes (the visible spectrum), and those longer still (such as infra-red rays) give the sensation of heat. Thus infra-red rays and microwaves are used for cooking.

LIGHT AND WAVES

Some baking is in fact done in microwave ovens, especially bread intended for a particularly long shelf life (e.g. 90 days). The bread is baked in a laminated plastic bag and is also irradiated so as to delay the development of mould. Wrapping of this kind must not be permeable (see Chapter 3); hence if the bag is punctured the long-life feature of the bread is completely lost.

Finally in the electromagnetic spectrum, mention must be made of television and radio waves whose characteristic property is that they can travel great distances through space. Even though we talk of short wave, medium wave and long wave radio it should be remembered that all radio waves have much longer wavelengths than other members of the electromagnetic spectrum. The range of wavelength covered in the whole spectrum is truly enormous; for example that of green light is about one half of a millionth of a metre (5×10^{-7} m), whilst that of a long wave radio station (Radio 4 in Great Britain) is 1500 metres.

Revision 10
1. Give three examples of luminous bodies and three of non-luminous bodies.
2. A picture of a white cottage with a red roof surrounded by green grass is viewed through a piece of red cellophane. How will it appear to be coloured?
3. State a law which is obeyed by a ray of light striking a mirror.
4. What is meant by 'critical angle'?
5. What is meant by 'plane polarised light'?
6. Describe two ways in which the strength of a sugar solution may be determined.

Chapter 11

Principles of Current Electricity

In the chemistry section of the book the structure of the atom will be discussed. To understand the phenomenon of electricity, it is necessary to know that each atom consists of a nucleus around which are spinning particles called electrons. These electrons are said to be particles of negative electricity and the nucleus in the free atom has in it positive electricity in amount equal to the negative electricity possessed by the spinning electrons.

If two substances, such as ebonite and fur, are rubbed together some of the electrons in the fur break away and get attached to the ebonite. The ebonite then has an excess of negative static electricity and the fur is left with excess positive static electricity. They are said to be charged negatively and positively respectively and each has the property of attracting light bodies such as tiny pieces of paper. A comb, after combing hair, will exhibit the same property.

If the ebonite, fur, comb and hair were examined with suitable apparatus, it would be found that the excess positive and negative electricity would not have spread itself throughout the substance, but would have remained at the places which were rubbed. Such bodies are called *insulators*.

If, however, the negative charge from an ebonite rod is rubbed on to a copper sphere held on a glass stand, the charge will distribute itself evenly over the copper surface. The copper is said to be a *conductor* as it allows electrons to flow in it. The glass stand will not take up any charge as it is an insulator and electrons will not readily move in it.

Electric wiring in a house is made of conducting material so that electrons can flow in it, whereas switches contain insulating material to stop the electrons flowing when the switch is in the 'off' position. The wires are covered with insulating material so that electrons do not escape. The electrons are sometimes made to flow by what is termed a 'cell'. An ordinary torch cell (sometimes called a battery) is suitable. If a torch bulb is connected to the terminals of the cell then it will light. This is because electrons are flowing from one terminal of the cell to the other through the bulb.

PRINCIPLES OF CURRENT ELECTRICITY

For diagrammatic purposes the cell is given the symbol indicated in Fig. 11.1 which is called a circuit diagram. Although it is the electrons which actually move, it is conventional to call the direction opposite to that in which the electrons flow the direction of flow of an *electric current*, i.e., from the positive terminal of the battery to the negative terminal, through the bulb. An electric current is defined as 'the rate at which electrons flow past a point in a circuit in the direction opposite to that in which the current is measured.' If the electrons flow quickly, the current is larger than if they flowed slowly. An instrument to measure current is called an ammeter. The cell provides the force necessary to drive electrons round the circuit. This *electromotive force*, as it is called, is measured in volts. An instrument to measure voltage is called a voltmeter.

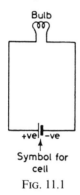

Fig. 11.1

If the cell in the above circuit were a 1.5-volt cell and if it were replaced by a 3-volt cell (twice as powerful), then it would send twice as many electrons round the circuit in a given time and would, therefore, double the current. This is one of the fundamental laws of electricity—the current flowing in a conductor is proportional to the voltage (or potential difference) applied to it. It is called Ohm's Law.

However, the voltage is not the only factor which governs the size of the current in a circuit. An analogy may help to explain this. If two equally powerful cars start off at the same place for the same destination and one driver chooses a narrow road while the other goes on a motorway, the second driver will be able to move more quickly. Similarly, electrons can flow more easily in thick wires. A piece of long thin wire introduced into a circuit of otherwise thick wires would have the same effect as a narrow bridge would have on a racing track—it would reduce the speed of all cars. The long narrow wire is called a *resistor* as it resists the flow of electrons.

Long narrow wires are high resistors and are used to cut down the current in a circuit, while thick wires which have a very low resistance (they are low resistors) are used for connecting wires.

Sometimes the current in an electric circuit has two or more paths from which to choose. A circuit containing two resistors A and B, giving a choice of path, can be drawn as in Fig. 11.2. A and B are said to be in parallel, and the current from the cell will divide itself between the two resistors, the larger current flowing through the smaller resistor as measured by the ammeters. If A and B are resistors of equal resistance, then the same current will flow through each. If B is twice as large as A then twice as much current will flow in A.

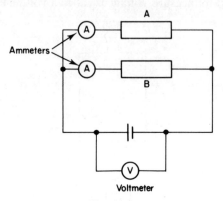

FIG. 11.2

Although Ohm's law states that the current flowing in a given conductor is proportional to the voltage applied across it, the size of the current for a given voltage will vary with different conductors, and, in fact, depends on the resistance of the conductor.

If the resistances are measured in units called ohms and current in units called amperes, then the voltage is equal to the product of current and resistance.

Consider the circuit given in Fig. 11.2. Suppose that a 2-volt cell is used. Then 2 volts are applied across each of the resistors. If A is 1 ohm in resistance and B is 2 ohms, then

$$2 \text{ (volts)} = 1 \text{ (ohm)} \times \text{current in A}$$
$$\text{and } 2 \text{ (volts)} = 2 \text{ (ohms)} \times \text{current in B}$$

The current in A is, therefore, 2 amperes (A) and that in B is 1 A. Hotplates in electric cookers are joined in parallel.

Ohm's law can be used to find the current in a circuit where the resistors are joined end to end or 'in series' rather than in parallel. A typical circuit diagram is that of Fig. 11.3 where A, B and C are three resistors of resistance 1, 2, and 3 ohms respectively. In this case the voltage is applied across all three resistors, so we add them together to give a total of 6 ohms. This can only be done if they are in series. Note that the voltmeter is in parallel with the cell.

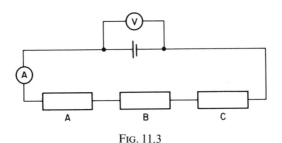

Fig. 11.3

If the battery is a 12-volt battery then:

$$12 = 6 \times \text{current in whole circuit}$$

Therefore, the current in the whole circuit is 2 A. This same current passes through all the resistors in a series circuit.

It is possible to obtain resistors whose value varies by moving a slider along a bar so that different lengths of thin resistance wire are included in a circuit.

When a current flows in a conductor, heat is given out. An example of this is an electric fire or oven. The heat given out each second depends on the applied voltage and the current. The product of volts applied and the current in amperes flowing through the heating element gives the heat energy given out per second in watts.

Energy per second (watts) = voltage × current (amperes)

Example:

A 1000 watt (1 kilowatt) electric fire is run from the 250-volt mains.

(a) What current passes?

volts × current in amperes = watts
250 × current in amperes = 1000

$$\text{Therefore, current} = \frac{1000}{250}$$
$$= 4 \text{ A}$$

(b) What is the resistance of the element?

volts = resistance in ohms × current in amperes
250 = resistance in ohms × 4

$$\text{Therefore, resistance} = \frac{250}{4}$$
$$= 62.5 \text{ ohms}.$$

As a safeguard against fire which could be caused by the overloading of wires, fuses are incorporated in circuits. These are often wires of low melting point. Thin diameters are chosen so that when the maximum current for the wiring is exceeded, the fuse wire becomes so hot that it melts and breaks. The current can then no longer flow. Fuses are usually embedded in asbestos or some other substance which is a good insulator of heat.

A two-bar fire is usually a 2 kilowatt fire with two 1 kW elements in parallel. It therefore takes 8 A and a three-bar, 3 kilowatt fire would take 12 A. A 3 kW fire can safely be run from a 13 A fused plug and socket but neither it nor a 2 kW one must be run from a plug protected by a 3 or 5 A fuse, as the fuse would melt and disconnect the fire.

For large electric ovens, the wires carrying the current have to be very thick to prevent overheating. They frequently consist of a number of strands bound together in one insulator as this is easier to install than one single thick wire, since it is more pliable.

The cost of electricity is easily calculated if the price of a unit is known. A unit is the amount of electricity that a 1 kilowatt element would use in 1 hour when run off the correct voltage. Thus a 30 kW oven run for 2 hours would use 60 kilowatt-hour units. At 3p a unit this would cost £1.80.

Very few people have not played with magnets at some time and their power to attract certain materials, notably iron and steel, is generally known. If a bar of soft iron has a piece of insulated (covered) wire coiled round it a number of times and a current is passed through the wire, the soft iron bar behaves like a magnet. It will attract other pieces of soft iron and iron filings will cling to its ends. Immediately the current is switched off, the iron bar ceases to be a magnet and iron filings clinging to it will drop away. It is said that the passage of an electric current has a magnetic effect called the *electromagnetic effect*, and a piece of soft iron with a coil of insulated wire around it is called an electromagnet.

Figure 11.4 is a suitable circuit for demonstrating the electromagnetic effect.

An electromagnet is the main feature of an electric bell (Fig. 11.5). The current from the dry cell enters at A, and C normally touches

PRINCIPLES OF CURRENT ELECTRICITY

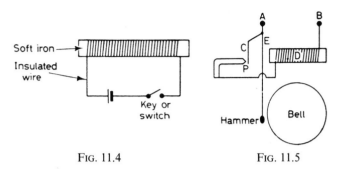

Fig. 11.4   Fig. 11.5

the contact point P so that current flows around the electromagnet, D, and back to the cell connected to B. However, as soon as the current flows, the electromagnet attracts E which is made of soft iron. When E moves towards D, the hammer hits the bell, but at the same time C is pulled away from P. This stops the current flowing through the circuit and D no longer attracts E, which as it is a spring, returns to its original position so that C makes contact with P again and the whole procedure is repeated several times a second, the bell being hit each time.

The larger the current in the wire surrounding an electromagnet, the greater its attraction for a piece of iron. If a piece of soft iron is held a fixed distance, say 1 cm, from the end of an electromagnet, then if a very small current is passed through the electromagnet the iron will not be attracted. If, however, the current is gradually increased, then at a certain value, the soft iron will be attracted. This fact is used in safety 'cut-out' devices. The device is so designed that at a certain current (considered to be the maximum safe current for the piece of apparatus being protected) an iron bar is attracted to an electromagnet through which the current passes. The movement of the iron bar operates a lever which switches the device to the 'off' position, thus cutting off the current.

The instrument will not allow current to pass again until an operator puts the switch back to the 'on' position after rectifying any fault. As it is easier to flick a switch than to mend a fuse, this device, while serving the same purpose as a fuse, is an easier (although initially more costly) means of preventing overloading. However, both the fuse and the 'cut-out' differ from the bimetallic strip described in Chapter 5 which cuts the current off when the instrument gets too hot but allows the current to pass again when the instrument has cooled down, without any attention.

In Chapter 6, the thermocouple is described. If the two junctions of a pair of different wires are kept at different temperatures, then a current will flow (see Fig. 6.5). This fact, in conjunction with an electromagnet, is used to operate yet another safety device, one that ensures that if the pilot light of a gas oven blows out, the gas is cut off.

One junction of the thermocouple is kept just above the pilot flame (Fig. 11.6), while the other is kept well away from it. As long as the flame is alight the junctions are at different temperatures and a current flows in the circuit which includes an electromagnet, A. B, which is made of soft iron, is attracted to the electromagnet and gas can enter the apparatus through the inlet hole C. However, if the flame is extinguished, the current no longer flows as both junctions are at about the same temperature, and B is no longer attracted to the electromagnet, and being on springs D, it drops back and closes the inlet hole for the gas.

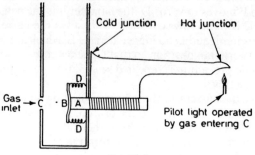

FIG.11.6

In the same way as insulation having become worn can cause two wires to touch, so it can cause a wire to touch the metal casing of an electrical machine or oven. If a worker then touches this casing, the electric current will pass through him to 'earth' and he may get a very unpleasant electric shock which could even be fatal. This is especially likely if he has wet hands which make the passage of current from casing to hands easier. For this reason all electrical appliances are 'earthed'. The earthing wire leads from the outer casing, sometimes through the plug and socket, to metal piping or plate embedded in the earth.

Most readers will be familiar with the fused plug used for equipment taking a current not exceeding 13A. The three terminals (see Fig. 11.7) are labelled L (live), N (neutral) and E (earth). In the UK the three wires from appliances are coded as follows: Brown to *live*, blue to *neutral* and green/yellow to *earth*. The incoming lead is securely gripped where it enters the plug, to prevent the wire becoming strained in use and so breaking at the terminals.

PRINCIPLES OF CURRENT ELECTRICITY 83

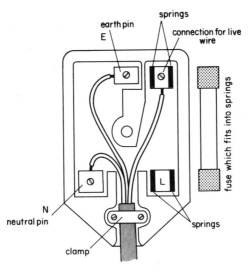

FIG. 11.7

So far we have considered electrons moving continuously round an electric circuit in one direction. This is called *direct current*, and the force which drives it is *direct voltage*. However, the national grid which supplies electricity to our homes and factories does not operate with direct voltage. Instead it supplies *alternating voltage* and so drives alternating current. An alternating current is so called because the direction of flow of the electrons of which it consists is regularly reversed so that it flows first in one direction and then in the other. The reversal rate or *alternating frequency* of the mains is measured as the number of complete cycles of alternation per second; one such cycle is called one hertz. In the UK this frequency is 50 hertz. The change in voltage in each cycle is smooth and a graph of voltage against time for 1/25th second is shown in Fig. 11.8.

The mains voltage in the UK changes from +320V to −320V 50 times each second. When any voltage is applied across a resistor a current flows through it so dissipating power. If it is great enough, the resistor may become warm. The heating effect of an alternating current with a peak voltage of 320V is about the same as that for a direct voltage of 240V. This is called the root mean square (rms) voltage. If the rms voltage is used in an Ohm's law calculation, the current thus found will be the rms current. The rms currents and voltages are the accepted ones when dealing with alternating currents and voltages since they can be used in calculations involving

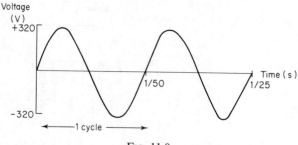

Fig. 11.8

resistance, power and energy. This is because they are based on equality of heat produced by direct currents and voltages.

Alternating current is used mainly because it is easier to transmit over large distances. Power stations generate at high potentials of thousands of volts and transformers are used to make the conversion to the required mains voltage of 240V. Transformers can only be used with alternating voltages.

Electrical power is used in bakeries to provide heat, for example in the various types of electric oven, although heating by gas or oil is also common. Power is also needed to mix dough, perhaps giving a measured quantity of work to the dough in a very short time. It is also necessary to transport the products of the bakery through proofers, ovens and coolers, perhaps at precisely controlled speeds and temperatures. Additionally, automatic dampers may be needed to give conditions to suit the type of bake required. All this implies that a large modern bakery presents a large demand for electricity. In such a case the alternating voltage provided may be from a *three-phase supply*.

Figure 11.9 shows three alternating voltages all of which have precisely the same frequency and reach the same peak voltage, but they reach these voltage peaks at different times. In the figure the voltage A commences at zero and reaches its positive peak after a quarter of one cycle. The voltages B and C follow at equal intervals of time before A reaches its positive peak again. A, B and C are said to be three phases of a single supply voltage and the supply providing them would be a three-phase supply. A typical system might operate using four wires. One is needed for each phase and one for the common return of the three circuits to which they are attached. The electric motor driving a mixer or providing the motive power for a band or rack oven may be so designed that it operates with a three-phase supply. Such motors are smaller and cheaper than would have been the case had they been designed for operation from a

PRINCIPLES OF CURRENT ELECTRICITY   85

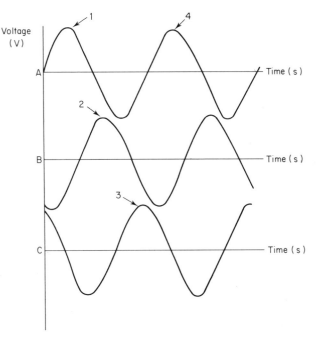

Fig. 11.9

single phase supply. Each phase supplies current and hence power to the motor. Provided the three phase currents have nearly equal rms values the current in the common return lead is very small.

Revision 11
1. Define the term 'electric current'.
2. In what units are the following measured—current, electromotive force, resistance?
3. State Ohm's law.
4. An electric light bulb is labelled 60W 240V. What is its resistance? How much would it cost to use it for 25 hours if the cost of electricity is 2 p per kilowatt-hour?
5. How would you demonstrate the electromagnetic effect?
6. A 10 ohm and a 15 ohm resistor are each connected in parallel across a 2V supply. Calculate (a) the current flowing in each resistor, (b) the total current flowing, and (c) the value of a single resistor which if connected instead would have allowed the same current to flow.

7. If two resistors R_1 and R_2 are connected in parallel across a supply voltage, V, show that the current flowing is the same as that through a resistance R across the same supply where $1/R = 1/R_1 + 1/R_2$.
8. Two lamps are respectively rated 75 W and 100 W at 240 V. Calculate the resistance of each and, assuming that these resistances remained constant, deduce the potential difference which would be developed across each lamp were they to be connected in series. Hence calculate the power dissipated in the series combination of the two lamps. Would they glow brighter or duller than before?
9. The resistance of metal filaments, and in particular that of metal lamp filaments, increases with temperature. In Question 8 would the current flowing through the series combination of lamps have been slightly more or less than would have been the case had (as assumed) the resistances remained constant? Give reasons for your answer.
10. Three pieces of electrical equipment in a bakery are intended to operate from a 240 V supply. They are rated at 2.5 kW, 1.25 kW and 500 W respectively. Each has to be protected by a fuse. 13 A, 5 A and 3 A fuses are available. Which fuses should be used?

PART TWO

CHEMISTRY

C. A. Russell

Chapter 12

Chemical Changes

All of life is marked by changes, and the most obvious of these are those taking place in the material world around us. Whether we are considering the climate, the position of a train on a journey, the appearance and behaviour ('properties') of substances that are heated or placed in new surroundings, or the growth of a human body, we are dealing with the world of matter. Of course, the list can be extended indefinitely.

Physicists and chemists are both interested in these changes, but by common consent they have divided the field between them. The physical changes dealt with by the physicist are those happenings that leave the substances involved much as they were so far as their essential nature is concerned. They include electrification, magnetisation, illumination, temperature changes, changes in position and so on. These are usually reversed easily and in none of them is any new substance formed.

The chemist, on the other hand, is concerned with those changes in which one substance is converted into another, or into several others. He *is* interested in the formation of new substances. Changes in which new kinds of matter are produced are called 'chemical changes' or 'chemical reactions'. We can, therefore, define chemical changes thus: A *chemical change* (or *reaction*) is a change resulting in the formation of new chemical substances.

There is no need to add anything to that definition, although it will be noted that, in a chemical change as thus defined, there will always be the disappearance of the properties of the substances at the beginning (reagents) and the appearance of new properties, those of the products.

We must not confuse the appearance of two substances with the change of a substance to merely a different physical form. Thus the combination of hydrogen and oxygen to form water is a true chemical reaction, but the melting of ice to water is not, because both ice and water are forms of the same substance and can be easily interconverted.

Thus, chemistry is about the conversion of materials into others. It is obviously not going to get very far unless it can offer some understanding of the nature of the substances that do react. The more we know about the players in the chemical game the more we shall be able to predict successfully the outcome. Hence chemists want to know not only how a given substance will behave in given circumstances. (e.g., flour in baking), but also of what it is made. This is the other great aspect of chemistry, the nature of materials. For this reason it is sometimes described as 'the science of substances'. We shall now see how these millions of different substances fall into three simple classes.

THREE CLASSES OF MATERIALS

Let us suppose our problem is to study the changes that go on during the baking of a loaf. The obvious starting-point is an examination of the chief raw material, flour.

The first conclusion is that flour varies from one sample to another. This suggests, perhaps, that our task is likely to be fairly complicated and that flour may be a mixture of several substances which, in different proportions, will give it a slightly different appearance.

This is confirmed by a number of simple tests as the following:

(a) Warm a little flour in a vessel connected to a vacuum pump; it will lose up to 15% of its weight, as that amount of water is liberated as vapour; if we take the product and put back an equal quantity of water to that which has been removed, the flour becomes much as it was at the beginning.

(b) Shake up the flour with cold water, filter off the insoluble part and evaporate the water; a small residue containing sugar and some other things remains.

(c) Repeat this process, but with alcohol (ethanol) instead of water; on evaporation, we have a different residue (proteins, etc.).

(d) Heat the flour on a crucible lid until no further changes seem to occur; after burning, it forms a black mass (carbon) which eventually burns away, leaving only a white ash (potassium phosphate, chiefly).

All these results suggest that flour contains quite a few different constituents. It does indeed contain water, sugar, proteins and potassium phosphate in addition to its chief component, starch. Separation such as we have described is fairly easy, and, except for the last test, can be reversed; i.e. the separated components can be

CHEMICAL CHANGES 91

put back without making the product very different from the flour at the start.

Suppose we now concentrate on any one of these separated substances, e.g., sugar. It will not take long to show that what we have called 'sugar' is actually made up of two main components known as sucrose and maltose. These can be separated, though with some difficulty. Let us suppose that this has been done and we now focus our efforts on to the substance sucrose (ordinary sugar). Here we find things are no longer what they were; it has been relatively straightforward to isolate the 'sugar' from the flour, then to separate the components of this. But to break the sucrose down further looks like being very difficult. Experiment shows that the task is impossible as long as we want to separate it into substances which will readily mix again with each other to re-form the original material.

For example we can treat it with concentrated sulphuric acid (a drastic procedure), and we obtain water and carbon, a black solid. But mixing these two does not form sucrose!

We are thus led to suppose that there is a difference between (say) a combination like sugar, starch, water, etc., to form flour and one between water and carbon to give sucrose. The first type can be easily and reversibly split into its components and is called a 'mixture'. In the second, there is some kind of force holding the water and carbon together which gives the product quite new properties; such a force we call a 'chemical force', and such a combination a 'compound'.

Compounds and mixtures differ in the following ways:

1. Compounds cannot be separated into their components by ordinary physical means, but mixtures can. Methods like dissolving out or boiling off one component are useless for compounds but may be satisfactory for mixtures.
2. When compounds are broken down into their constituents, they cannot usually be re-formed from these constituents merely by mixing them. That is not true of mixtures.
3. Compounds do not show the properties of their constituents, but mixtures do. Flour shows properties of starch, proteins, sugars, etc., but table sugar does not show any resemblance to water (a liquid) or carbon (a black insoluble solid).
4. Compounds have a definite composition by weight. We know, for instance, that in sucrose, the ratio of water to carbon is always 1.00 to 0.725. This fact about compounds is sometimes called the Law of Constant Composition, or the Law of Definite Proportions. But it does not apply to mixtures which can have any composition whatsoever.

The basic reason for all these differences is that, in compounds, there are very powerful chemical forces holding the components together and that these forces have far-reaching effects on the properties of the compounds they form. Such chemical linking is subject to certain definite laws and this accounts for the paragraph above.

To return to our investigation of flour, we concluded that it was a mixture of certain ingredients, including sucrose. This seemed to be a compound of carbon and water. We may now enquire whether our process of breaking down, or analysis, has finished. Can we break down carbon and water into anything simpler? Experiment shows that carbon cannot be split into anything simpler but water can; the latter is itself a compound of hydrogen and oxygen (*see also* p. 150) and these in their turn resist all attempts at further decomposition. This brings us to a third category of substances, those that cannot be split into anything simpler. These are termed elements. They are the ultimate units of which all matter is composed and there are well over 100 of them. These can combine in millions of different ways, and the products are the compounds. These in their turn can be mixed with each other or with elements to form an infinite number of possible mixtures. It is this total of mixtures, compounds and elements that confronts us as the material universe.

This section is concluded by summarising the information discussed about flour, then by giving formal definitions of the three terms.

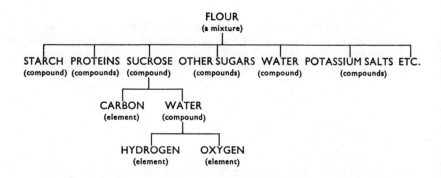

The definitions are:
Element—a substance that cannot be decomposed into any simpler substance.
Compound—a substance consisting of two or more elements chemically joined in definite proportions by weight.

CHEMICAL CHANGES 93

Mixture—any aggregation of two or more components that are not chemically joined in definite proportions by weight.

Revision 12.1
1. Do you think the following changes are chemical or physical? Give reasons: (a) The burning of coal, (b) the glowing of an electric light filament, (c) the action of water on sugar, (d) the action of heat on sugar, (e) the bleaching of flour, (f) the milling of flour.
2. Define a compound. Summarise in tabular form its chief differences from a mixture.

SPEEDING UP CHEMICAL REACTIONS

Some reactions are so fast that they seem to be instantaneous—for example, the neutralisation of an acid by an alkali or an explosion of flour-dust in air. Others are so slow that they seem not to be taking place at all. Fortunately, however, the rates of chemical reactions can be controlled to some extent. There are two factors which affect the speeds of chemical changes: temperature and catalysts.

1. TEMPERATURE
A simple rule holds good here: the higher the temperature the faster the reaction.

Hence, to accelerate reactions that are slow at room temperature we apply heat. This includes reactions that are so low at ordinary temperatures that they are never noticed until heat is applied. It is partly for this reason that foodstuffs are often cooked; desirable chemical changes take place fairly quickly at oven temperatures. In baking, for example, the liberation of carbon dioxide either by yeast or baking-powder occurs at a reasonable rate. There are other reasons for cooking which are physical and not chemical, but the latter must not be forgotten.

Conversely, reactions that take place too quickly at room temperature can be slowed down by cooling below that level. Refrigeration is for precisely this purpose. The undesired reactions are those which occur as food 'goes bad' and produces unpleasant or poisonous by-products. Perishable foods that are particularly susceptible to this deterioration are, of course, things such as milk, eggs, butter and meat. The alternative is to keep out the substances that cause reaction and, if these include air, it is possible to gain protection by use of airtight containers. But refrigeration is often far more convenient. A deep-freeze is more effective than a domestic refrigerator, operating at $-18°C$ ($0°F$) or cooler, while the main compartment of a refrigerator rarely falls below the freezing-point of water, $0°C$ ($32°F$).

As a rough guide, we may say that the rates of chemical reactions in general are doubled for every $10°C$ (or $18°F$) rise in temperature.

One other point may be mentioned here. Many chemical reactions give out heat; they are termed 'exothermic'. When one occurs, therefore, the temperature will get higher as the reaction proceeds, unless measures are taken to conduct away the heat evolved. Thus, the reaction becomes faster and faster until all the reagents are consumed. This is what happens when a fire becomes out of control. It is for this reason that air and gas supplies in an oven have to be regulated carefully, for if the supply of reagents is restricted the reaction cannot proceed and so the temperature remains at the desired level.

2. Catalysts

There are certain substances that will exert a marked effect on the rate of a chemical reaction; they are called catalysts, and are best defined thus:

Catalyst—a substance that alters the rate of a chemical reaction without itself being chemically changed by it.

In the preparation of oxygen in the laboratory from potassium chlorate (p. 127) manganese (IV) oxide is used to accelerate the reaction. In the industrial formation of ammonia from hydrogen and nitrogen a mixture of metal oxides is used for the same purpose. Many other examples exist of simple catalysts used to hasten a chemical reaction.

Sometimes a catalyst is used to slow down a reaction. It is then called an 'inhibitor'. Such substances are added to hydrogen peroxide, for example, to prevent its rapid decomposition into water and oxygen, and others are employed in foodstuffs to prevent rapid deterioration.

The most important catalysts for our purposes are those which operate in the chemistry of natural products and are called 'enzymes'.

Enzymes—complex organic molecules produced by living cells for catalysis of reactions necessary to the organisms concerned.

It is important to note that enzymes are not themselves alive. They are just very complicated molecules (proteins) secreted by living organisms.

All catalysts are limited to certain reactions and have no action on others. We say their action is highly specific. This is as true of simple inorganic catalysts such as manganese (II) oxide as it is of the enzymes. Table 12.1 shows important enzymes and the reactions

CHEMICAL CHANGES

TABLE 12.1

Enzyme	Occurrence	Reaction	Page
diastase	malt	starch → maltose	211
amylase	saliva, pancreas	starch → maltose	211
invertase	yeast, intestine	sucrose → invert sugar	134, 209
maltase	yeast, intestine	maltose → glucose	210
zymase	yeast	glucose → alcohol + carbon dioxide	134, 142, 209
trypsin	duodenum	protein → amino-acids etc.	216
lipase	pancreas, stomach	fats → glycerol + acids etc.	215
rennin	stomach, rennet	caseinogen → casein	104

they catalyse; it also includes reference to their occurrence and to the pages where their action is further discussed.

Revision 12.2
1. Define the terms catalyst and enzyme.
2. Give two standard methods for preservation of 'perishable' foodstuffs, with the reasons for using them.

Chapter 13

Solutions and Colloids

Everyone knows the kind of thing that happens when a lump of sugar is added to a cup of water; all the solid disappears, and its place is taken by a liquid which appears to be homogeneous, i.e., the same all through. There is a resemblance to the behaviour of a lump of fat placed on a warm surface, but the two processes are not the same thing. The sugar dissolves, the fat melts. In melting, only one kind of substance is present at the start, e.g., solid fat. But in dissolving, there are at least two phases at the beginning, phases being systems with bounding surfaces between them; usually there is a solid and a liquid phase at first, and a single liquid phase after dissolving has taken place. Although products from both processes are homogeneous, in melting only one kind of particle is present, but in a solution at least two must be there.

We start with a few definitions:

Solution—a homogeneous mixture.

Solvent—that component of a solution having the same physical state as the solution itself; therefore in a liquid solution (the most common), the solvent is the liquid component; where both components are liquid it is a matter of choice as to which is called solvent, often this name being given to the component present in larger quantity. The solvent 'does the dissolving'.

Solute—the dissolved substance in a solution; the component(s) other than the solvent.

Although solvents are most usually thought of as being liquids, they can have any physical form, as Table 13.1 suggests:

TABLE 13.1

Solvent Phase	Solute Phase	Some Typical Examples		
		Solution	Solvent	Solute
gas	gas	air	nitrogen	oxygen, etc.
liquid	gas	soda-water	water	carbon dioxide
liquid	liquid	vinegar	water	acetic acid, etc.
liquid	solid	syrup	water	sugars, etc.
solid	solid	brass	copper	zinc

In the discussion that follows, however, attention will be concentrated almost entirely on the solution in liquid solvents.

SOLUBILITY

One of the most obvious things about many solutions is that one cannot go on adding solute to a fixed volume of solvent indefinitely; there comes a time when no more will dissolve and the remainder merely falls to the bottom of the vessel. This maximum concentration of a solution is termed the solubility of the solute, commonly expressed in g/litre, though $g\,dm^{-3}$ would be preferable in the SI system. It will vary with temperature, usually increasing as the solvent becomes hotter. Solubility depends very much on the nature of the solvent; substances that are very soluble in water are generally insoluble in ether, and vice versa; but in similar solvents, such as methanol and ethanol, solubilities are usually similar.

Where the solute is also a liquid, one may have an infinite solubility, where the two liquids are miscible in all proportions, or a limited solubility beyond which further addition of solute merely causes separation of a second non-miscible layer. With gases and solids in liquid solvents solubility is always limited, and in some cases may be so small that it is immeasurable. In all probability there is no such thing as a completely insoluble material.

When a solution contains an amount of solute corresponding to the solubility of the latter, it is said to be saturated; if it has less, it is unsaturated. The real test is whether it changes composition when in contact with the solid material; a saturated solution is unchanged in these circumstances. This is the best way of looking at saturation, for in certain circumstances a solution may have more solute than the solubility indicates, i.e., more than when it is saturated; but this cannot happen in the presence of solid material. A 'supersaturated' solution of this kind is formed when a saturated solution cools (when the solubility drops) or when it evaporates (so increasing the ratio solute/solvent). Such a solution is not stable and, in contact with a crystal of the solid, or on shaking, or in contact with dust or other solid particles it will deposit the solute in excess of the amount to give a saturated solution.

It will now be convenient to give a few more definitions:
Solubility—the extent to which a substance will dissolve in a given solvent; usually the number of grams dissolved in the presence of excess solid in a fixed volume of solution (e.g., 1 cubic decimetre or 1 litre).
Saturated solution—a solution that remains unchanged in contact with solid solute.

Unsaturated solution—a solution that dissolves further solute when placed in contact with it.

Supersaturated solution—a solution containing more dissolved solute than the saturated solution.

DEPOSITION OF SOLID FROM SOLUTION

When a solid is deposited from a solution as the latter cools or evaporates, its particles often assume a characteristic geometrical form. Such particles are called crystals, and the process is known as crystallisation. There are several common shapes that crystals can take, such as a cube (like salt) or a rhombohedron (like quartz). When several substances have the same crystalline shape they are said to be 'isomorphous'.

Crystallisation from solution is a valuable method of purifying dissolved solids; if impurities are more soluble than the solute they will tend to remain in solution after the latter has crystallised; if they are less soluble they will precipitate out first and can be filtered off before the main deposition of solid. This is the method used to purify most crystalline materials of the food industries such as sugar, salt and the solid acids referred to on pp. 175–6. The process is usually repeated several times and is then called 'recrystallisation'.

It is also worth nothing that the longer a crystal takes to form the bigger it will generally be. Thus rapid evaporation of a salt solution, for example, will produce large numbers of tiny crystals; but if evaporation is allowed to take place only over a long period perhaps only one or two large crystals will form.

SOME COMMON SOLVENTS

Water—will dissolve most substances in at least small quantities; dissolves most acids, and many salts, also all alkalis. Does not dissolve many organic substances, important exceptions being alcohol, acids and sugars.

Ether—very inflammable, but capable of dissolving a vast range of organic substances, though not sugars, starch or cellulose.

Carbon tetrachloride—non-inflammable, this is exceptionally efficient in dissolving fats and grease generally; has been used for dry-cleaning, but its vapours are rather toxic.

Trichlorethylene—similar in action to carbon tetrachloride, but less poisonous. Used in degreasing metals and in dry-cleaning.

Acetone—unlike the last three examples this solvent mixes with water; it dissolves a rather wider range of solutes than the others, and is inflammable.

Alcohol (ethanol)—also miscible with water and inflammable, this dissolves many organic substances, though in some cases less

SOLUTIONS AND COLLOIDS 99

readily than the other organic solvents mentioned. It will act as a solvent for some inorganic substances as well.

Revision 13.1
1. Name three solvents for fats.
2. Explain the terms, saturated solution and supersaturated solution.

COLLOIDS

THE NATURE OF COLLOIDS
In 1861 it was discovered that certain substances in solution (or apparently so) failed to pass easily through a parchment membrane, whereas others readily diffused through. These two classes were termed respectively 'colloids' and 'crystalloids'. The word 'colloid' means 'glue-like', glue being one of the substances behaving in this way, while 'crystalloid' was the name given to the other group, most of which existed in crystalline forms.

It is not now customary to speak of substances as being colloids or crystalloids, as many can exist in both conditions at different times. It is better to speak of the 'colloidal state' rather than 'colloidal substances'; sometimes the phrase 'colloidal solution' is used.

The first thing to grasp about colloids is that, strictly, they are not true solutions at all. A true solution is a one-phase system, a colloid is not. Colloids consist of two phases, and are, therefore, heterogeneous, not homogeneous. These two phases are:

1. The disperse phase: this corresponds to the solute, and consists of minute particles whose diameters are *roughly* between the limits of 10^{-6} and 10^{-9} m.
2. The dispersion medium, or continuous phase: equivalent of the solvent: this is the medium in which the colloidal particles are dispersed.

This immediately raises the question: How do colloids differ from suspensions? Broadly, the answer is in size and stability. In a suspension the particles are big enough to be seen, if not with the naked eye, at least with the microscope, and that means that they must be no smaller than 0.2×10^{-6} m diameter. On the other hand, colloidal particles cannot be seen even through a conventional microscope, any more than those in a true solution. They must therefore be smaller than about 10^{-6} m across, but cannot be as small as those of a true solution or they would be able to pass through the small 'holes' in the parchment membranes.

Thus, in terms of size, colloidal particles are midway between those of true solutions and true suspensions. But there is another factor which differentiates them from suspended particles. In a suspension, there are two phases and the particles of the suspended phase tend gradually to fall to the bottom of the vessel under the influence of gravity—hence the injunction on many bottles of medicine to 'shake the bottle' before pouring out the suspension. With a colloidal system, however, the particles of the disperse phase tend to remain uniformly distributed for an indefinite period; the colloid is said to be 'stabilised' and, in this way, it differs from a suspension.

We are now fairly clear why colloids remain stable. There must be some kind of repulsion between the particles of the disperse phase which prevents them from coming together, or coagulating. This repulsion is believed to be electrical in nature, all particles of this phase bearing the same kind of electrical charge. As we know, like charges repel. This charge can only come from the dispersion medium, and the important point is that ions are absorbed from the latter by the particles of the disperse phase; this will, of course, mean that the two phases will have opposite charges. Suppose positive ions are absorbed from the originally neutral dispersion medium; then all the particles of the disperse phase will become coated with a positively charged sheath, while the other phase will become negative. It may be illustrated very simply as shown in Fig. 13.1.

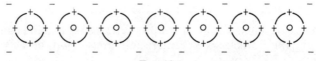

Fig. 13.1

A convincing piece of evidence for this comes in the phenomenon of electrophoresis. The apparatus illustrated in Fig. 13.2 can be used. When a current is passed, the particles of the disperse phase may move towards one electrode, and will therefore bear a charge opposite in sign to that of the electrode.

CLASSIFICATION OF COLLOIDS

Colloids may be classified into many groups according to the physical states of the two phases involved. There are, for example, colloidal systems known as aerosols in which solid or liquid particles are suspended in gases, and this includes many fogs, mists and of course the well-known aerosol sprays. The commonest

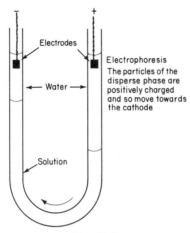

Fig. 13.2

systems, however, and those of most direct concern to us, are those in which the continuous phase is a liquid, and these are sometimes termed 'sols'.

Sols are themselves divided into two broad classes: *Lyophilic* (= 'liquid loving')—on cooling or evaporation these give rise to solids which can be readily re-converted to a sol by adding more liquid; *Lyophobic* (= 'liquid-hating')—those which under these conditions do not readily re-form the sols.

We can summarise the main differences as shown in Table 13.2 for sols with water as the continuous phase.

TABLE 13.2

Properties	Lyophilic sols	Lyophobic sols
1. Viscosity as compared with that of water	Much higher	About the same
2. Effect of electrophoresis	Migration in either direction or not at all	Migration in one direction only
3. Ease of coagulation	Coagulated only with difficulty	Easily coagulated

Lyophobic sols

Colloidal systems of this kind are not of great importance in the processes of bakery. Consequently we shall have little to say about these. It is perhaps sufficient to point out that they are prepared with some difficulty and coagulated with ease. On addition of small quantities of electrolytes the dispersed particles are precipitated out; i.e. the sol is coagulated. The reason appears to be that the colloidal particles absorb ions opposite in charge to their own. Hence the total charge on the particles is reduced and so their mutual repulsion becomes less. They tend to come together, and the sol coagulates.

Examples of this kind of system include colloidal solutions of metals in water, and some of the milky-looking products obtained by dispersing sulphur and sulphides in it.

Lyophilic sols

These are much easier to obtain than the lyophobic type; in many cases all that is necessary is to warm or grind the two components together. As we have seen, they are quite stable and even if the continuous phase is removed by evaporation, the resultant solid can be reconverted into the colloidal state merely by adding more liquid.

From this it seems that there must be a stabilising factor that is not present in lyophobic systems—at least not to the same degree. This factor is believed to be *hydration*. Each of the particles is believed to be surrounded by water molecules which form a sheath around it and thereby prevent it from coalescing with similar particles. Hydration probably operates in conjunction with the mutual repulsion due to like charges on the particles, and thus reinforces their effects.

A lyophilic sol is obtained when water is added to flour; insoluble proteins, e.g. glutenin, take on a colloidal form, and this continues until the gluten is fully hydrated, at which point the mixture ceases to be sticky. The colloidal protein appears to form a spongy network which forms the skeleton of the dough and which is rendered elastic by another protein, gliadin, which is soluble. This network helps to retain the gas set free during baking, and also has embedded in it granules of starch.

On baking, the hydrated protein coagulates. At the same time, a process known as denaturation also occurs, in which the long protein molecules tend to become uncoiled.

Gels

When a colloid sets to a semi-solid with no liquid apparently present the product is called a gel. The most familiar example is the setting of table jellies, whose chief component is the protein gelatin. Recently alginate gels (from seaweed) have become common in foodstuffs, e.g., as texture improvers for sponge cakes. The simplest explanation is that the solid particles unite to form a three-dimensional mesh in the pores of which the liquid is retained, rather as water is kept in the pores of a sponge.

Gels are formed from starch when it is heated with water. The starch granules start to swell at about 60°C and by 70°C a paste is formed which is essentially a gel, and which on dilution with hot water gives the mis-named 'starch solution'. This is the preparation once used for starching clothes. Now, starch is also present in flour and it may be asked if gelation takes place there on heating; the answer seems to be that to some extent it does, but does not go to completion because the temperature is not high enough for a sufficiently long time, and also the amount of water is not enough.

Revision 13.2

1. Define the terms 'disperse phase', 'continuous phase', 'lyophilic sol', 'lyophobic sol'.
2. What two factors help to stabilise a protein sol?
3. Give a brief account of colloidal systems met with in bakery processes.

EMULSIONS

An emulsion is a two-phase system in which one liquid (the disperse phase) is suspended as minute droplets in a second liquid (the continuous phase). It is probably not a true colloid as the size of the suspended particles is rather too great, though they are usually far too small to be seen with the naked eye. The two immiscible liquids do not settle out into separate layers so long as they are truly 'emulsified'; instead the system takes on a milky, opalescent appearance, and to the naked eye *seems* to be the same all through, i.e., homogeneous. It is in fact, however, a two-phase system.

Now suppose that water and paraffin are shaken together. It is possible that an emulsion may be formed, but unlikely that it will persist for long; separation into two layers will probably occur very speedily. Such systems are known as 'temporary emulsions'. An example that is sometimes quoted is milk with a high proportion of cream; a sample of this on standing does tend to form two layers, with the cream rising to the top. The lower layer, richer in water and poorer in fats, is an emulsion in its own right, however, and this will not settle into any further layers, so this watery fluid can hardly be called a 'temporary emulsion'.

Emulsions which do not settle out easily are termed 'permanent', though there is no sharp distinction between the two groups. Except in the case of very dilute emulsions, all permanent stable systems are found to contain small amounts of a third component which directly contributes to their stability and which is called an 'emulsifier' or 'emulsifying agent'. Soaps and detergents are common emulsifying agents; their cleansing action is in part due to their ability to form a stable emulsion of the grease (dirt) in water, so enabling it to be washed away. A similar kind of emulsification occurs in the process of digestion, when partial breakdown of fats occurs in the body, and one of the products is a kind of soap which then emulsifies the rest of the fat. The emulsified fat is then capable of being absorbed in the walls of the intestine. One of the reasons for the ready assimilation of milk is the fact that it is already emulsified. In the baking process emulsifiers play a vital part in helping to retain moisture, promote aeration and inhibit staling. They have other functions as well and are usually glycerides or related substances (p. 215)

The emulsifying agent in milk, caseinogen, belongs to a different class: proteins. It enables the suspended droplets of fat to remain emulsified in water, although it is destroyed by the substance rennet which converts it into the insoluble protein casein. This reaction takes place in the stomach and can be induced artificially by adding rennet to milk; it is used in the preparation of junkets and of cheese, when the curd formed consists of casein which is associated with some of the fat.

Emulsions may be 'broken' (i.e., converted into two layers) by removal of the emulsifying agent, or by rapid movement as in the churning of milk into butter.

Revision 13.3
1. What is an emulsion?
2. Why is milk sometimes called a 'temporary emulsion'?

Chapter 14

Atoms and Molecules

It was on August 6, 1945, at Hiroshima, Japan, that the first atomic bomb in warfare was exploded. The results are now part of history, for, within hours, Japan was suing for peace, and the worst war man has known was over. With this event a new era had begun, not only politically but also in a scientific sense. The atomic bomb is generally regarded as having been the herald of the 'Atomic Age', and certainly the possibilities of almost limitless quantities of energy being obtainable from the atom became common knowledge for the first time.

If it is true that the key to the production of power in the future lies in harnessing the atom, it is equally certain that the idea of an atom lies at the heart of all modern theories of chemistry and of many physical theories. This has been true for many years and long before popular attention was called to atomic energy by the events of 1945. In fact, the atomic theory that is used in simple chemistry dates back to the early 1800s, to the work of John Dalton.

With our greatly increased store of knowledge we cannot accept as always and completely true all that Dalton suggested; but for many simple purposes his theory is a most valuable starting point. Indeed, it is true to say that, unless the main points of Dalton's theory are clearly understood, it will be impossible to grasp the chemistry necessary to bakery science (or any other kind, for that matter).

Thus, if we wish to understand something of the processes that occur when the various operations of bakery are performed, we have to use the same basic ideas as those necessary to explain the workings of modern weapons of destruction and of nuclear power stations and these are the theories of the atom. We shall look first at what atoms are, then how they combine, and finally at methods of representing their combination in chemical equations.

NATURE AND CHARACTERISTICS OF ATOMS

The simplest definition of an *atom* is the smallest part of an element that can exist.

Suppose we are interested in the composition of any substance e.g., flour. The first task would be to separate this mixture into its constituents, then to try to break down any compounds that were present (as starch) into their elements. If this has been done, we shall have a complete analysis, or breakdown, of the material into its basic components, such as carbon, hydrogen, oxygen and so on.

It may be, however, that we want to know more about these elements which, as we have seen, cannot be decomposed into anything simpler. Since, therefore, we shall never obtain any other kind of substance from them, the only thing to do is to examine the units of which they consist. This will presuppose that such units do exist, and this has not always been believed. There have been those who held that matter is infinitely divisible and continuous; there will never be a point beyond which further division into yet smaller parts will not be possible. But the chemist believes that the process of breakdown into ever smaller particles must have a limit, and the particles at this limit he calls 'atoms' (from the Greek for 'uncuttable' or 'unsplittable'). In view of this it is rather odd that the phrase 'splitting the atom', which contains an implicit contradiction, should be so often used in popular speech.

We see, therefore, that all elements are made up of atoms, or at least that is what chemists believe to be the case. There is a great deal of evidence, but most of it will not be touched on here. If the point is granted, however, it immediately follows that all matter is atomic, since all matter is made up of combinations of the 100 or so elements. It is now desirable to see what are the characteristics of these atoms, and for a start we shall look at the 'Dalton' atom, as from many points of view this description of an atom is as useful as any for a chemist. At the same time we shall notice some of the ways in which modern discoveries have modified these original ideas. The following, then, are some of the more important ideas about atoms that Dalton suggested in his Atomic Theory:

1. *Matter is not infinitely divisible:* i.e., one cannot keep on dividing and dividing a quantity of any substance indefinitely. At last one comes to particles which resist all attempts to break them down, and these are the 'atoms'. Originally Dalton used this word for the ultimate particles of elements and compounds; nowadays, we confine the term 'atom' to elements only, and call the smallest particles of compounds 'molecules', but more of this point later. Apart from the question of names, however, we need to ask if Dalton's idea is still acceptable: we find that it requires a slight modification. It is certainly correct to say that an atom exists and is the smallest particle of an element that can exist; but it does not follow from this that it cannot be split into something different from the element.

In fact, this is just what can happen. For all atoms are themselves composed of electrical charges, positive and negative, and, by subjecting most common kinds of atoms to tremendous forces, we can break them down into atoms of other elements which are simpler in structure; in this process electrical particles will be liberated as well, together with huge amounts of energy. This is what occurred in the 1945 atomic bomb. In some cases the process of atomic breakdown takes place spontaneously, as the atoms seem to be inherently unstable, and this natural process is referred to as 'radioactivity'. It is now possible to effect the opposite of atomic fission, atomic *fusion*, in which simpler atoms come together to form atoms of a more complex kind, again with evolution of enormous energies.

In view of all this, it is easy to adopt the superficial view that Dalton's atomic theory is useless: nothing could be further from the truth, however. Atomic disintegration is not a factor normally to be reckoned with in chemistry, or in those chemical and physical changes that constitute our everyday experience. For these the basic postulates of Dalton's theory are as applicable as they ever were and, although it is an approximation, the deviations from Dalton's theory are too small or too rare to justify its abandonment as an ordinary working theory. As for the definition of an atom, we can modify the one above by saying that 'an atom is the smallest particle of an element still to retain the properties of that element'. Thus if an atom of uranium does break down into other particles these will not be atoms of uranium, and will not have the original properties. They will be either electrical particles or atoms of a different element, and this brings us to a second aspect of Dalton's theory.

2. *Atoms are of different kinds:* in Dalton's own words, there is 'a considerable number of ... elemental principles'. In this he made what was probably his most original contribution. Previously men (the Greeks, for instance) had believed in atoms, but all atoms had been regarded as of the same kind; the difference between, say, tin and zinc was only in different patterns of the same kind of atoms. But Dalton said the atoms of tin were different from those of zinc, and so on. That the atoms of one element differ from those of all the other elements is a statement that chemists have held ever since Dalton's theory was accepted.

3. *Atoms of one substance are all alike:* in modern terminology we should say atoms of each element, and molecules of each compound, are all alike; each atom of sulphur is like every other atom of sulphur, and every molecule of water is identical with every other water molecule. In so far as this applies to ordinary chemical behaviour, this is very nearly true, but again it is not completely correct. We now know that an element can have several different kinds of

atom termed 'isotopes', the principal difference between them being weight and those properties like density, rate of diffusion, etc., that depend directly on weight of particles. Chemically, isotopes may for present purposes be regarded as behaving in identical ways.

It so happens that, in nature, any specimen of an element (or a compound for that matter) is almost exactly the same as any other. Exceptions to this are so rare that we shall ignore them. The existence of isotopes demands some explanation for this and it is found that the relative amounts of each isotope in any specimen are very nearly constant. In other words the isotopes have been so well mixed that the composition of an element is fixed within very fine limits, and ordinary analytical methods cannot detect a difference in specimens that have been obtained by a great variety of methods.

Since each isotope has a definite weight, and since the ratio of isotopes is constant, there will be an average weight for an atom of every element, and this is defined as 'atomic weight', i.e., the number of times an atom of an element is heavier than one atom of hydrogen.† We express it this way (and not in grams or other units of mass) because the actual weight is so small and because we are chiefly interested in relative weights. Thus it follows that every element has a definite atomic weight, this being an average of the weights of each of its naturally occurring isotopes, the exact value depending on their proportion. The atomic weights of the common elements are given in the appendix, and are of the greatest importance. Table 14.1 gives a list of a few elements showing also the weights of the natural isotopes, the most abundant being printed in bold type.

TABLE 14.1

Element	Approximate atomic weight	Weights of natural isotopes
Chlorine	35.5	**35**, 37
Potassium	39.1	**39**, 40, 41
Argon	39.9	36, 38, **40**
Carbon	12.0	**12**, 13, 14

†A more accurate definition refers the weights to $\frac{1}{12}$ of the weight of a carbon isotope; this gives fewer fractional values but makes the atomic weight for hydrogen 1.007 97, with oxygen 15.999 4. For approximate work, the two scales become the same. It is also common to speak of atomic mass and molecular mass (instead of atomic and molecular weights). In this book, however, we have deliberately retained the older terms as being more natural and convenient.

It is interesting to note that Dalton himself did not think the absolute weights of atoms could be determined but that relative values could be easily obtained. In fact, he was wrong on this first point (as he was on his methods for determining atomic weights), but in all chemical work we measure relative weights and these are all we normally find necessary.

4. Chemical combination occurs between simple numbers of atoms: this is the point that most clearly links the atomic theory with simple experimental facts. These facts are that, when elements combine and other chemical reactions take place, there is always a very simple relationship between the weights of substances taking part. Without going into details we can infer from this that the most probable explanation is that the atoms combine in quite simple ways with each other, in ratios like 1:2, 1:3, 1:4, 2:3, 3:5, and so on. The only exceptions to this are unimportant, and we can take it as a perfectly valid generalisation.

This immediately means that compounds will have definite, and often simple, compositions, determined by the number of atoms of each component.

SYMBOLS AND FORMULAE

To most people chemical formulae represent simply a method used by chemists for making their subject harder than it need be—a kind of code known only to the initiated! In fact, however, a chemical formula is merely a neat and concise method of conveying a lot of meaning in a small space, employing very simple rules. First, we shall have to look at:

Atomic symbols: Every element is given a symbol of one or two letters, and this atomic symbol stands for one atom of the element (and not, strictly speaking, the element in general). The symbol is derived often from the English name for the element; as there are only 26 letters in our alphabet we cannot represent all 110 elements by a single letter and so some have two letters derived from the English name; others are based on Latin words or on pseudo-Latin names derived from the Arabic. Table 14.2 lists some of the symbols of the elements; a fuller list will be found on p. 219.

Symbols such as these date not from Dalton but a little later, and are due to the Swedish chemist J. J. Berzelius. They replace the mysterious signs used up to that time. Combinations of these symbols lead to:

Molecular formulae: Let us assume that we know the composition of a number of compounds, and in particular the number of the

TABLE 14.2

Name	Symbol	Origin of symbol
argon	A	English name
aluminium	Al	English name
carbon	C	English name
calcium	Ca	English name
chlorine	Cl	English name
cobalt	Co	English name
fluorine	F	English name
hydrogen	H	English name
iodine	I	English name
magnesium	Mg	English name
manganese	Mn	English name
nitrogen	N	English name
nickel	Ni	English name
oxygen	O	English name
phosphorus	P	English name
sulphur	S	English name
zinc	Zn	English name
silver	Ag	Latin name (argentum)
gold	Au	Latin name (aurum)
copper	Cu	Latin name (cuprum)
iron	Fe	Latin name (ferrum)
mercury	Hg	Latin name (hydrargentum)
lead	Pb	Latin name (plumbum)
antimony	Sb	Latin name (stibium)
tin	Sn	Latin name (stannum)
potassium	K	Arabic name (kalium)
sodium	Na	Arabic name (natron)

different atoms in each molecule. The formula of the molecule is then derived as follows. The symbol of each atom present is written down, and is followed by a number indicating how many atoms of the element concerned are in the molecule; this number is written slightly below the line of the rest of the print and is omitted when it is only one. If a group of atoms appears more than once in a molecule this group (or 'radical') is placed inside brackets and the suffix immediately after the closing bracket applies to everything within the brackets. When it is desired to represent more than one molecule of the same kind, a number is prefixed to the formula of the molecule, but this time placed on the line of print. Some examples will make

ATOMS AND MOLECULES 111

these points clear:

CuO: one molecule of copper oxide, containing one atom each of copper and oxygen;
H_2O: one molecule of water, consisting of two atoms of hydrogen and one of oxygen;
H_2SO_4: one molecule of sulphuric acid, consisting of two atoms of hydrogen, one of sulphur and four of oxygen;
$2CO_2$: two molecules of carbon dioxide, each consisting of one atom of carbon and two of oxygen;
$3O_2$: three molecules of oxygen, each containing two atoms;
A: one molecule of argon consisting of one atom;
$Al(NO_3)_3$: one molecule of aluminium nitrate consisting of one aluminium atom and three nitrate groups, each of which comprises one nitrogen atom and three oxygen atoms;
$5(NH_4)_2SO_4$: five molecules of ammonium sulphate, each of which contains two ammonium radicals (each having one nitrogen and four hydrogen atoms) and one atom of sulphur and four of oxygen.

In this way it becomes easy to see from any molecular formula exactly what it conveys. There are certain conventions that determine the order in which we write down the atoms in a molecule. These need not detain use here, but it will be noted that metals and hydrogen always come first in simple formulae. Organic compounds present special problems here and we shall see some of these in Chapter 21.

Molecular weights: The molecular weight of a substance is simply the sum of the individual atomic weights, regard being paid to the number of atoms concerned. In the examples following, the atomic weights used are: $H = 1$, $C = 12$, $O = 16$, $N = 14$, $S = 32$.

CH_4 molecular weight $= 12 + 4 = 16$
CO_2 molecular weight $= 12 + (2 \times 16) = 44$
NH_3 molecular weight $= 14 + 3 = 17$
H_2SO_4 molecular weight $= 2 + 32 + (4 \times 16) = 98$
$(NH_4)_2SO_4$ molecular weight $= 2(14 + 4) + 32 + (4 \times 16) = 132$
$C_{12}H_{22}O_{11}$ molecular weight $= (12 \times 12) + 22 + (11 \times 16) = 342$

Like atomic weights these figures are not absolute weights but ratios. It is often common to speak of the molecular weight in grams, e.g., 44 g of CO_2. This unit is called the *mole* and is of great value in many chemical calculations, especially volumetric analysis. It is helpful to think of it as that amount of substance containing as many particles (molecules, ions or atoms—whichever are being specified) as there are atoms of hydrogen in 1 g of that element; this needs slight modification for the latest atomic weight scale but is quite adequate for ordinary use.

The molecular formula of a substance tells us only what is there, and gives no indication *how* the various atoms are linked to each other. Indeed, it was at one time doubted whether this could be found out at all! However, the nature of chemical bonds has now been much examined and we have a good idea as to how atoms are linked. Knowledge of this explains why the formulae are what they are; why water is H_2O and not H_3O or HO_5 for example. We shall, therefore, look at the subject of valency.

VALENCY

This word has two rather different uses. It may mean, in general, the forces holding atoms together in a molecule; or it may have a more precise meaning and denote the number of atoms of hydrogen that it should be possible to unite with one atom of the element in question. It is in this latter sense that we often say that 'an atom of oxygen has two valencies', or 'the valency of oxygen is two', or just 'oxygen is divalent'. Some elements have more than one possible valency state, but will not usually change from one to the other unless acted on by an oxidising or reducing agent (p. 188). In such cases of 'polyvalency' the different valency states are sometimes indicated by different names; for metals, lower states used to be implied by the ending '-ous' and higher by the ending '-ic'. More recently it has become the practice to incorporate the valency number into the name; thus CuO is copper (II) oxide (instead of cupric) and Cu_2O is copper(I) oxide (instead of cuprous).

All of this is of the greatest importance for writing formulae because there is a simple rule connecting these with valencies: when a molecule is formed the atoms unite in such numbers that all valency bonds are utilised. A molecule can be regarded as being formed by union of all chemical bonds that each atom can have, the number of these being indicated by the numerical value of its 'valency'. Hence our principle implies that each of these bonds must be paired off, and none must be 'spare'.

A view of atoms as minute spheres having 'hooks' on their surface which have at all costs to be linked to similar features on other atoms is highly fanciful, but also highly successful in conveying in simple terms how molecules are composed. It is exactly how many mid-Victorian chemists depicted molecules, and (provided we recognise its limitations) is a good place to start. Thus, if we take the valencies of hydrogen, oxygen and carbon as one, two and four respectively we can write the following combinations of these elements:

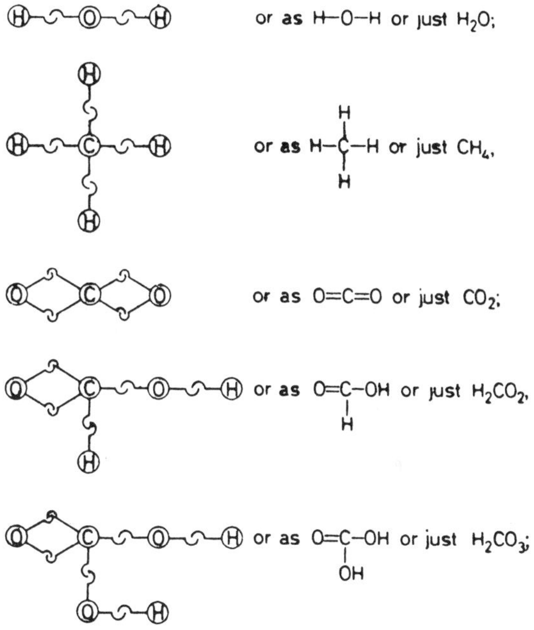

H–O–H or just H_2O;

or as H–C–H or just CH_4, (with H above and below)

or as O=C=O or just CO_2;

or as O=C–OH or just H_2CO_2, (with H below C)

or as O=C–OH or just H_2CO_3; (with OH below C)

It happens that many compounds are built up from two atoms, and for these we can arrive at a formula particularly simply. Suppose we combine an element A, valency x with an element B, valency y; for all valencies to be satisfied we could have y atoms of A and x atoms of B, each having a possibility of forming xy bonds and therefore satisfying each other. Hence the formula will be A_yB_x, unless x and y have a common factor in which case we divide both by that. Here are some illustrations:

aluminium (valency 3) and oxygen (valency 2): Al_2O_3
aluminium (valency 3) and chlorine (valency 1): $AlCl_3$
aluminium (valency 3) and nitrogen (valency 3): AlN (*not* Al_3N_3)
chlorine (valency 1) and oxygen (valency 2): Cl_2O

Mention has already been made of *radicals*. These are groups of atoms which tend to remain intact through many chemical changes, and which are recognisable as distinct entities. These radicals are never found uncombined as they all have 'spare' or 'free' valencies, but when they are found in compounds they are combined in just the same way as single atoms. As will be seen from the examples below, the atoms *within* the radicals obey the usual rules of valency.

Table 14.3 gives the valencies of the chief elements and radicals:

TABLE 14.3

1. Elements

Valency	Metals	Non-metals
0	—	A, He
1	Ag, Cu (I), K, Na	Br, Cl, H, I
2	Ca, Cu (II), Fe (II), Hg, Mg, Pb, Sn (II), Zn	O, S
3	Al, Cr, Fe (III)	N, P
4	Sn (IV), Ti	C, S, Si
5	V	N, P
6	Cr, U	S

2. Radicals

Valency	Radicals		Structure
1	OH	hydroxide,	—O—H
	NO_2	nitrite,	—O—N=O
	HCO_3	bicarbonate,	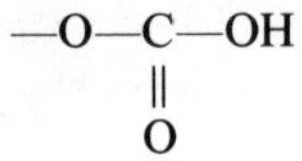 —O—C(=O)—OH
2	CO_3	carbonate,	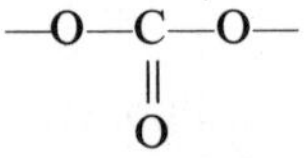 —O—C(=O)—O—
	SO_4	sulphate,	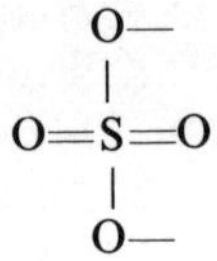 O=S(=O)(—O—)—O—
	SO_3	sulphite,	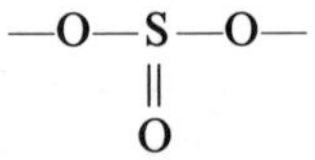 —O—S(=O)—O—
3	PO_4	phosphate,	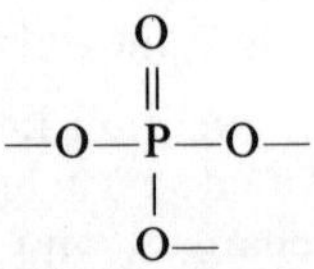 —O—P(=O)(—O—)—O—

Of these atoms and radicals, the zero-valent elements (the inert gases) form no compounds; the metals combine with non-metals and

ATOMS AND MOLECULES

radicals; the non-metals combine with each other and the metals; the radicals are found in combination with hydrogen or metals. There is one radical, ammonium or NH_4, which quite exceptionally behaves like metals in its combinations. It has a valency of one.

At this point the question may be raised 'Why do atoms have definite valencies?'. A full answer to this is beyond the scope of this book, but we can give a brief indication of the cause of valency in general.

It will be recalled that the atom is electrical in nature and, in fact, it consists of two parts. There is a positively charged *nucleus*, each element having atoms bearing identical charges on this nucleus. If the atom is electrically neutral, the nucleus may be pictured as being surrounded by a number of electrons that will equal the positive charge it bears; since electrons are negative the net effect is to give a neutral atom. The electrons may be pictured as encircling the nucleus as the planets do the sun (only with the difference that the atomic particles are charged). Now it happens that not all arrangements of electrons are stable and, in fact, only in the inert gases does this occur spontaneously. In all other elements the atoms have to modify their natural arrangements to gain stability, and this is why an element will undergo a chemical reaction. An electronically unstable element will give rise to a stable compound.

The precise way in which this can happen will vary, but the case of sodium and chlorine illustrates the simplest type of change. A neutral sodium atom has 11 electrons, one more than the stable number of 10; a neutral chlorine atom, on the other hand, with 17 electrons lacks one for the desired 18. Clearly, therefore, stability in both could be achieved if the 'surplus' electron were to be transferred from the sodium to the chlorine, and this is what happens in practice. The product is sodium chloride, or common salt. But in common salt we do not meet *uncharged* sodium or chlorine atoms, for, in the transfer of this electron, they have become charged, the sodium positively and the chlorine negatively, and these charged atoms we call ions. In positive ions the charge on the nucleus exceeds that of all electrons, and vice versa for negative ions. For sodium and chlorine the situations can be summed up thus:

	Sodium atom	Sodium ion	Chlorine atom	Chlorine ion
Charge on nucleus	+11	+11	+17	+17
Number of electrons	−11	−10	+17	−18
Net charge of atom	0	+1	0	−1
Conventional symbol	Na	Na^+	Cl	Cl^-

116   PHYSICS AND CHEMISTRY OF BAKING

That kind of compound we call *electrovalent* or *ionic*. Many metal salts and oxides are similarly constituted, and consist of vast assemblages (or crystals) of positive and negative ions all kept in position by electrostatic forces and very hard to dislodge. Such compounds are solids with high melting temperatures; once dissolved in water or melted, however, the ions move freely and may be detected by electrolysis. The capacity to be an electrolyte is the most fundamental characteristic of electrovalent compounds.

Let us now consider another type of substance, hydrogen gas. Hydrogen atoms are exceptional in that they are 'satisfied' with either 0 or 2 electrons around the nucleus; the neutral atom has just one. Its state of instability is relieved by the simple process of allying itself to another lone atom, the two of them *sharing* their pool of 2 electrons, thus:

$$H:H$$

There is no transfer as in electrovalency, no ions exist and there are no strong forces between molecules. Hence such compounds cannot be electrolysed and are often gases or liquids. The type of valency exemplified here is *covalent* bonding. Some examples follow where crosses represent electrons from one kind of atom, and dots electrons from the other kind; in each case we are concerned with the outermost electrons only. The neutral atoms are shown first.

H×	·Cl:	H×Cl:	H−Cl
H×	·C·	H×C×H (with H above and below)	H−C−H (with H above and below)
:O	×C×	:O×C×O:	O=C=O
×N×	·Cl:	:Cl×N×Cl: with :Cl: above	Cl−N−Cl with Cl above

From these examples we can see the reasons for a number of facts certainly difficult to explain as simply on any other theory. They include the following:

(a) *The nature of valency:* in the case of common salt and many other compounds, the force holding the atoms together is simply the electrostatic attraction between ions bearing opposite charges. This is obviously a better explanation than 'hooks' sticking out of atoms! Not all compounds, as we have seen, are held together in this way, but all chemical union does involve a modification of the electronic structure of the atoms concerned.

(b) *The numerical value of the valency:* sodium and chlorine are both monovalent because one electron only is involved in each case to accomplish electronically stable states. Calcium, on the other hand, has two electrons surplus to the stable number and, when it combines with chlorine, two atoms of the latter are required to accept these two electrons; hence the product here is $CaCl_2$, and calcium is divalent. Always the value of the valency bears a direct relation to the number of electrons more or less than the preferred number. In the examples here it is just that, though, with polyvalent elements especially, other factors have to be taken into account. At least, however, we can see why each element has its own definite valencies, as for each only a small number of stable arrangements for the electrons exists and in many cases there is only one.

(c) *Differences between compounds and their components:* Why should salt be so different from sodium (a shiny reactive metal) and chlorine (a yellow poisonous gas)? The answer to this is simply that the free elements exist as uncharged atoms (or molecules as Cl_2) but salt consists of their ions. That is to say, the free and combined atoms have different electronic arrangements and, as many physical and all chemical properties depend on the electronic structures, free elements and their compounds are completely different in their appearance, reactions and other characteristics.

(d) *Fact of electrolysis:* this is the decomposition of certain liquids by the passage of an electric current through them. For instance, pure water does not conduct anything more than a minute electric current, but if common salt is added it becomes a good conductor, and is decomposed at the same time into hydrogen and oxygen. The only explanation is that charged particles must now be present and that these will carry the current from one electrode, i.e., the plate where the current enters or leaves, to the other. On arrival, they lose their charges and become free atoms which are intensely reactive and at once initiate a series of chemical changes ultimately resulting in the formation of hydrogen and oxygen. Again it must be emphasised that not all compounds are capable of acting thus; only those compounds which are held together by attraction between ions can act as 'electrolytes'. Nevertheless this does include a large number of important substances (acids, bases and salts).

Thus our simple hypothesis of direct transfer or sharing of electrons does explain many features of the chemistry of common salt and similar substances. It is, of course, too simple as it stands, and many compounds are formed by electronic changes rather different from this. But the essential point is the universal and vital connection between valency and atomic structure.

Where ions are formed (in the so-called 'ionic compounds') they exist in all conditions of the compound. But when the latter is melted or dissolved they become free to move; the compound is not ionised by becoming liquid, it is that already; but its ions become mobile, and this is often indicated by statements like this:

$$NaCl = Na^+ + Cl^-$$

Substances forming positive ions are hydrogen, metals and ammonium; all other common ions are negative, and include the rest of the radicals on p. 115 and also chloride (Cl^-), bromide (Br^-), and iodide (I^-). The number of charges on any ion equals the valency of that ion. Some more examples of ionisations follow:

$$CaCl_2 = Ca^{++} + 2Cl^-$$
$$K_2CO_3 = 2K^+ + CO_3^{--}$$
$$HNO_3 = H^+ + NO_3^-$$
$$(NH_4)_2SO_4 = 2NH_4^+ + SO_4^{--}$$
$$AlBr_3 = Al^{+++} + 3Br^-$$
$$CuSO_4 = Cu^{++} + SO_4^{--}$$

EQUATIONS

We have seen that formulae are merely convenient shorthand ways of representing the composition of molecules. It is clearly desirable to have a similarly simple device for summarising the course of chemical reactions, and such a device exists in the chemical equation. This is a way of writing down, in terms of molecular formulae, the reacting molecules and the products, with indications of the relative quantities of each. We write these on the smallest possible scale, however, showing the fewest numbers of molecules of each kind to obtain a complete statement of all that is involved. The expression is termed an 'equation' because equal numbers of atoms of each kind must appear on each side. This follows from the law that matter cannot be created or destroyed. Hence every atom that takes part must finish in one of the products.

To illustrate some of these points, we take the following equation which represents what happens when hydrogen burns in oxygen:

$$2H_2 + O_2 = 2H_2O$$

ATOMS AND MOLECULES

This means that *two molecules of hydrogen, each consisting of two atoms, combine with one molecule of oxygen, also containing two atoms, to form two molecules of water, each consisting of two atoms of hydrogen and one of oxygen.* Clearly something is wanted to avoid statements of that length! The above is the correct version of this equation; the following are all wrong, for various reasons:

$$2H_2 + O_2 = H_2O$$
$$H_2 + O_2 = H_2O$$

Neither of these 'balances'; i.e. there are more atoms one side than the other.

$$H_2 + O = H_2O$$

This does balance, but involves a wrong formula; the element oxygen is usually met in the form O_2; its molecule is diatomic not monatomic as this implies.

$$4H_2 + 2O_2 = 4H_2O$$

This equation balances and contains no wrong formulae, but it is unnecessarily complicated. Division throughout by 2 will give the first (correct) equation, and no purpose is served by writing more molecules than necessary.

Two points remain for discussion: What information is, and is not, conveyed by a chemical equation? With regard to the first, it can be said that a chemical equation tells us the number of molecules taking part and being formed. This information, coupled with a knowledge of atomic weights, enables us to predict the actual weights of each reagent and each product. This is the most obvious way in which atomic theory is related to simple experimental facts.

The equation will tell us the ratio in which molecules react. Knowledge of the molecular weight of each enables us to predict what will be the ratio of their actual weights. Thus suppose we have the reaction:

$$aA + bB = cC + dD,$$

and the molecular weights of A, B, C and D are respectively α, β, γ and δ. Then the ratio of A to B by weight will be $a\alpha/b\beta$, and that of C to D will be $c\gamma/d\delta$. Thus if we have, say, 5 g of A we shall need $5b\beta/a\alpha$ g of B, and shall obtain $5c\gamma/a\alpha$ g of C and $5d\delta/a\alpha$ g of D. Some specific examples below will amplify these points:

Example 14.1
Given that 44 g of carbon dioxide occupy $22.4\,dm^3$, calculate the weight of baking powder ($NaHCO_3$) required to produce $1\,dm^3$ on heating.

The equation is:
$$2NaHCO_3 = Na_2CO_3 + H_2O + CO_2$$
$$(2 \times 84) \quad (106) \quad (18) \quad (44)$$

The figures in brackets are the appropriate molecular weights. From these we can say that

168 g $NaHCO_3$ give 44 g, or 22.4 dm^3, CO_2

$\dfrac{168}{22.4}$ g $NaHCO_3$ give 1 dm^3, of CO_2

i.e., 7.5 g of baking powder.

Example 14.2
Calculate the weight of ammonium chloride needed to produce 10 g of 'agene' (NCl_3), formerly used as a bleach for flour.

The equation is:
$$3Cl_2 + NH_4Cl = NCl_3 + 4HCl$$
$$(6 \times 35.5) \quad 53.5 \quad (120.5) \quad (4 \times 36.5)$$

Hence:

120.5 g 'agene' are obtained from $\quad$ 53.5 g NH_4Cl

1 g 'agene' is obtained from $\quad \dfrac{53.5}{120.5}$ g NH_4Cl

10 g 'agene' are obtained from $\dfrac{53.5 \times 10}{120.5}$ g NH_4Cl

i.e. 4.44 g of ammonium chloride.

Example 14.3
What weight of sodium carbonate is necessary to remove the hardness in water due to 5.4 g of calcium sulphate?

The equation is:
$$CaSO_4 + Na_2CO_3 = CaCO_3 \downarrow + Na_2SO_4$$
$$(136) \quad (106) \quad (100) \quad (142)$$

Thus:

136 g $CaSO_4$ requires $\quad$ 106 g Na_2CO_3

Therefore, 1 g $CaSO_4$ requires $\quad \dfrac{106}{136}$ g Na_2CO_3

Therefore, 5.4 g $CaSO_4$ requires $\dfrac{5.4 \times 106}{136}$ g Na_2CO_3

i.e. 4.2 g of sodium carbonate.

This kind of problem would be quite insoluble without the aid of the equation, and this indeed is its chief function. Sometimes equations may be modified slightly to convey more information. In place of the 'equal sign' a reversible arrow is used to indicate that the reaction can be reversed easily. Thus, instead of writing the two equations

$$CaO + CO_2 = CaCO_3$$

and

$$CaCO_3 = CaO + CO_2$$

we can write,

$$CaCO_3 \rightleftharpoons CaO + CO_2$$

because both reactions can be made to occur.

The formation of a precipitate from a clear solution is sometimes shown by a downward pointing arrow, and the evolution of a gas by an arrow pointing upwards, e.g.,

$$CaSO_4 + Na_2CO_3 = CaCO_3 \downarrow + Na_2SO_4$$
$$C_6H_{12}O_6 = 2C_2H_6O + 2CO_2 \uparrow$$

Despite the value of equations, it is possible to overestimate this, and it is worth recording what equations do not tell. Most important of all, they do not show whether the reaction they purport to represent can ever occur or not: also, they do not give information about temperature, pressure, catalyst or other necessary conditions. Finally, they do not afford any description of the reaction, whether it is fast or slow, explosive or gentle in action, and so on. But for the knowledge they do convey they are an indispensable part of chemistry.

Chapter 15

Air: Its Components

Modern chemistry began when people discovered the connection between two familiar things: the burning of fires and that invisible fluid we rather take for granted and which we call the atmosphere or air. Of course, it had been obvious for a long time that a connection between these two existed, but exactly what it was remained uncertain until less than 200 years ago. Most people were familiar with the effect of rapidly moving air (a wind or draught) on a fire, but it was not at all clear just why this was so.

To come as quickly as possible to an understanding of the nature of air, we shall make a number of simple statements or propositions and discuss each in turn. The order is not that in which the facts were discovered historically, but is probably the most logical arrangement.

1. Air is a substance: If we define a 'substance' or 'matter' as that which has weight (see p. 2), it can easily be shown that air fulfils that condition and therefore is material, or a substance. All that is necessary is to weigh a container with some air in it, draw out the air with a pump and weigh the container again (really 'empty' this time). It will be found that the air had a definite weight.

2. Air is necessary for combustion: 'Combustion' is the technical name for any kind of fire or burning. It cannot occur in the absence of air, except in a very few quite odd cases. This can be shown in many ways. One method is to place a lighted candle in a bell-jar (p. 124), and gradually pump out the air from the apparatus; the candle is rapidly extinguished. Alternatively magnesium may be heated indefinitely under a small pile of sand, and it will not catch fire despite its ability to do this with free access of air.

Since air has this property of allowing things to burn in it, it is said to 'support combustion', and this ability is shared by very few other gases. Closely related to this is its power to 'support respiration', for breathing of animals, and sometimes of plants, is a kind of controlled combustion of organic matter to produce energy, including heat.

3. Air becomes combined in combustion: As soon as one accepts the fact that combustion cannot occur without air, a new question arises. When a fire burns, is something removed from the air or added to it? To make matters simpler we can consider as a 'fire' the burning of a metal to an ash, e.g., magnesium.

In a case like this, one thing is clear from the start: only a limited quantity of metal can burn in a given amount of air. If a large weight of magnesium is heated in a little air, the burning soon stops and unchanged magnesium is left, together with some ash. This could be either because the metal has given something to the air which has become saturated after a while, or because the air has given something to the metal and in time all it had to give becomes used.

The first view looks attractive: something does seem to go into the atmosphere during combustion. It has one fatal weakness and that is that the weight of the ash is greater than the weight of the original metal; hence something must have been removed from the air and transferred to the metal. This can be demonstrated also by weighing the air before and after combustion, provided a closed vessel is used. It is then found that the gain in weight of the metal equals the loss in weight of the air. So a simple transfer from air to metal looks likely.

In other cases of combustion, however, the opposite may seem to be true. A lump of coal or coke, for example, burns away to a much smaller weight. The reason for this, however, is that the products here happen to be gaseous, not solids as with metals. If apparatus can be devised whereby all gaseous as well as solid products can be weighed, it will be found that there is an overall increase in weight in the starting materials, and an equal decrease for the air.

The modern theory of combustion thus acknowledges that the role played by air is to give something up to the burning body, and that this transfer may be examined by changes in weight.

4. Only part of the air becomes used in combustion: An odd fact about combustion is that when it has finished, and there is still some unburnt material present, there is always an appreciable amount of the air left, and it is impossible to continue burning in this. The apparatus shown in Fig. 15.1 can be used to demonstrate this. The bell-jar is supported in such a way that water can flow freely in and out. The air inside is sealed off by the water; only that which is originally there can be used for burning the candle which floats in

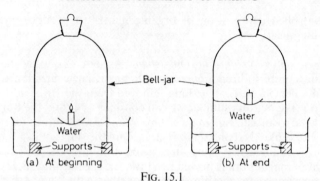

Fig. 15.1

some kind of dish. The candle may be lit with the rays from a burning glass, or can be burning before the apparatus is assembled. As soon as combustion begins, air is used, and so the total amount decreases causing the level of the water in the bell-jar to rise. This continues until the level is about one-fifth of the total distance from the original water level to the top of the jar. At this point combustion ceases and cannot be restarted; the 'air' that is left has none of the ability of fresh air to support burning (or respiration).

We are thus led to conclude that air, in fact, contains more than one ingredient. There is the part that supports combustion, and this is shown by accurate experiments to be 20.95% of the total volume; the rest is chemically inactive.

5. *Air is a mixture:* The fact that air is only partly involved in combustion could mean two things. On one hand it is possible that two (or more) ingredients of air are chemically joined together in the form of a compound which becomes decomposed by the burning bodies; on the other hand these components could be simply present as a mixture. Now we have already seen (p. 91) that there are certain fundamental differences between compounds and mixtures, and certain tests that can be used to distinguish between them. If we apply these tests, as we shall now do, we shall see that all the evidence points to air being a mixture. The active component is called oxygen, and the remainder consists of several gases, with nitrogen as the most important. The following facts are evidence for air being a mixture:

(a) *Variable composition:* compounds always have a fixed composition by weight, but mixtures do not. In the case of air the composition by weight (or volume) is very nearly constant, but not quite. At high altitudes, for instance, the proportion of oxygen is a little less than at sea-level.

AIR: ITS COMPONENTS 125

(b) *Separation by physical methods:* when clean, dry air is compressed to very high pressures and cooled to very low temperatures, it condenses to a blue liquid, 'liquid air'. If the temperature of this is now gradually raised, the different components become gases at different temperatures, and so can be separated. This process (fractional distillation of liquid air), being purely physical in nature, could only succeed if air were a mixture. If it were a compound it would all boil at the same temperature. It can be separated by other physical methods, as diffusion through porous containers, etc., but fractional distillation happens to be the most effective, and is used industrially.

(c) *Properties of its components:* in a mixture the properties of the constituents are always observable, but in a compound these usually disappear. In fact, air does show the properties of both oxygen and nitrogen, and has no properties not possessed by one or other of its constituents.

(d) *'Synthetic air':* if all the components of the atmosphere are mixed in the correct proportions at ordinary temperatures, when chemical combination can be ruled out, the product is indistinguishable from ordinary air. This must therefore be a mixture also.

To summarise what has been said so far, it may be stated that air consists of a physical mixture of about one-fifth oxygen and the remainder nitrogen and other gases; the oxygen alone is removed in combustion, and this process cannot normally take place unless oxygen is present. It is this gas which gives to air its unusual ability for supporting combustion and respiration.

We must now examine the components of the atmosphere in more detail.

CONSTITUENTS OF AIR

The following figures relate to a typical specimen of pure dry air, and indicate the proportions by volume of each constituent:

Nitrogen	78.08%
Oxygen	20.95%
Carbon dioxide	0.03%
The 'inert gases' (chiefly argon)	0.93%

Ordinary air also contains water vapour, the percentage of which depends on the atmospheric pressure. Suspended matter, such as dust particles, spores of yeast, etc., and water droplets may also be found. In the neighbourhood of industrial plant, other materials are likely, particularly soot and sulphur dioxide, and near the coast minute particles of common salt.

Revision 15.1

1. Explain why sand, water and carbon dioxide can all be used as fire extinguishers.
2. Does the *solid* material increase or decrease in weight when the following are heated in air: sugar, flour, iron, coke? Explain your answer.
3. On what evidence do we say that air is a mixture?
4. Explain why increase in the draught may raise the temperature of a fire. Can you suggest why too strong a wind may extinguish a fire?
5. Predict what would happen if the air in the apparatus in Fig. 15.1 were replaced by either oxygen or nitrogen. What would occur if free entry of water into the bell-jar were restricted, e.g., by placing the jar on the base of the container?

The four major constituents will now be considered in turn.

OXYGEN

The constituent of air that is responsible for most of its reactions is known as oxygen. This is, of course, an element, and in the air it occurs as such, i.e., without being combined with any other element. As oxygen is divalent it would not be expected to exist as single atoms, and it is usually met as the molecule O_2. Thus its molecular weight is twice its atomic weight, namely, 32.

It is, perhaps, worth mentioning that there is even more oxygen in a combined form in the earth's crust than there is in the air. It exists here as a great variety of compounds as oxides, carbonates, sulphates, etc., and altogether makes up nearly 50% of the crust of the earth.

PREPARATION
On an industrial scale, oxygen is obtained by the fractional distillation of liquid air. After a preliminary purification, air is liquefied by compression and cooling, and the oxygen separated from the other components by allowing the liquid to boil. Nitrogen tends to evaporate first (boiling point = 77 K), oxygen to come off next (boiling point = 90 K), and the other constituents boil at much higher temperatures. In fact, matters are rather more complicated than this simple explanation would suggest, but this is the essence of the process.

AIR: ITS COMPONENTS

Laboratory methods are not often needed, as pure oxygen can be purchased so easily. Fractionation of liquid air requires apparatus of too cumbersome a nature for this to be workable on a laboratory scale, so the gas has to be obtained from some source containing combined oxygen. There are scores of such sources that could be used but we shall be content with two of the simplest, both requiring the same apparatus. The reactions involved are:

1. Action of heat on mercury (II) oxide
Mercury (II) oxide (HgO) decomposes into its elements quite easily on warming:

$$2HgO = 2Hg + O_2$$

The reaction was important in that it played an important part in the work of Scheele, Priestley and Lavoisier in the eighteenth century through which oxygen was recognised as an element and the foundations of modern chemistry laid.

2. Catalysed decomposition of potassium chlorate
Potassium chlorate ($KClO_3$) decomposes, when heated, into potassium chloride and oxygen:

$$2KClO_3 = 2KCl + 3O_2$$

The reaction is made much easier by the addition of some manganese (IV) oxide (MnO_2). This does not enter the equation, as it remains at the end of the reaction and is not consumed. Its effect is to accelerate the above reaction and it can be recovered afterwards chemically unchanged. A substance behaving thus is said to be a catalyst (see p. 94).

We are now in a position to consider the apparatus used for the laboratory preparation of oxygen, which (as has been said) is identical for both reactions. In Fig. 15.2 we have a cross-sectional view of this (the usual way of depicting chemical equipment of this kind). The omission of supports for the boiling tube is quite intentional and indeed is customary; this is because the precise nature of such supports does not matter much, and diagrams should be as simple as possible. Other points that are quite general are that corks are not always shaded where the glass tube passes through, and all major items are indicated by appropriate labels.

The boiling tube is made of 'hard' glass, i.e., capable of withstanding a moderate amount of heat without softening. Notice that there is an air-space above the solid throughout its entire length; otherwise rapid formation of large volumes of gas at the left-hand end would force the solid up the tube, possibly blocking the exit.

128 PHYSICS AND CHEMISTRY OF BAKING

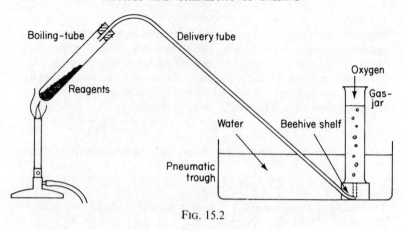

Fig. 15.2

The delivery tube is simply a glass tube bent to deliver gas under the beehive shelf. The latter is indicated in perspective in Fig. 15.2 (a) and is usually made of earthenware. It is constructed to act as a stand upon which an inverted gas-jar can rest, with an entry for the delivery tube and a hole through which the gas can enter the jar. The gas-jar (glass) is rather like an elongated jam-jar, but with a ground-glass flange at the open end upon which a slightly greased flat glass disc or cover can be slid to form a gas-tight seal.

In this experiment, the jar is upside down and is initially filled with water. (This is done by filling with water, placing on the cover, inverting and not removing the cover until the mouth of the jar is under the surface of the water.) Water is kept in the jar by the pressure of the atmosphere so long as no gas is present also (*cf.* the barometer). As soon as a bubble of gas emerges through the hole in the shelf into the jar it rises to the top and so displaces some of the liquid into the trough, and this is allowed to continue until the jar is full of gas and empty of water. The lid is placed under the jar and the whole assembly removed from the trough, another jar being introduced as before.

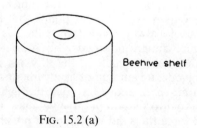

Fig. 15.2 (a)

AIR: ITS COMPONENTS

Physical Properties
Physical properties usually mean those which do not involve obviously chemical reactions. Oxygen is a colourless, odourless, tasteless gas. It is, of course, not merely non-poisonous but essential to animal life (respiration). It is slightly soluble in cold water; this is enough to supply the needs of fish which breathe the dissolved oxygen, but not sufficient to cause serious losses when collected over water as in its preparation.

Chemical Properties
The chief point here is that oxygen is the active constituent of air, and that most things that air will do oxygen does far more rapidly, e.g.

1. Combustion of elements

Most elements will burn in oxygen, some on strong heating only. Argon, chlorine, gold, and platinum are among the few that resist attack by this element at all temperatures. The rest form oxides, i.e., compounds with oxygen and one other element only. Some form more than one oxide, the product depending on conditions. A few examples are given in Table 15.1.

TABLE 15.1

Equation	Conditions
$C + O_2 = CO_2$	Strong heat; plenty of oxygen.
$2C + O_2 = 2CO$	Strong heat; oxygen supply restricted.
$2Ca + O_2 = 2CaO$	Slowly at room temperature.
$2Mg + O_2 = 2MgO$	Burns brilliantly on warming.
$S + O_2 = SO_2$	Moderate heating.
$3Fe + 2O_2 = Fe_3O_4$	Red heat.
$N_2 + O_2 = 2NO$	Very high temperatures, as in lightning flash.

Some metals do not burn well in oxygen but do corrode rapidly in air. This question is deferred till later (p. 188) but does provide an interesting exception to the rule that air is less active than oxygen.

2. Combustion of compounds

Many chemical compounds containing elements that are themselves inflammable will burn in oxygen, as will some that contain less than the maximum amount of oxygen, e.g.,

$$2H_2S + 3O_2 = 2H_2O + 2SO_2$$

The most important examples, however, concern the burning of organic (carbon) compounds. Nearly all of these contain hydrogen also, and the products of combustion are carbon dioxide and water, if there is enough oxygen. This includes the burning of hydrocarbons in fuel oils: e.g.,

$$2C_8H_{18} + 25O_2 = 16CO_2 + 18H_2O$$

combustion of alcohol as in methylated spirit:

$$C_2H_6O + 3O_2 = 2CO_2 + 3H_2O$$

the ignition of flour dust when mixed with air (the chief constituent being starch, a compound of formula $(C_6H_{10}O_5)_n$ where n is large but variable), and a vast number of other examples.

3. Other chemical reactions

In addition to the reactions involving quite high temperatures, there are a few others where oxygen produces chemical changes at around room temperature. The most interesting of these is its action on an alkaline solution of an organic compound known as pyrogallol; this will absorb all the oxygen in a vessel, combining with it to form new compounds. This occurs at room temperature and is a convenient way of estimating how much oxygen is present in a sample of gas, e.g., air. Other substances are known that behave similarly.

We cannot include the rusting or corrosion of metals as a property of pure oxygen because these involve the presence of other substances as well (see p. 188).

Uses: Pure oxygen is readily available compressed in metal cylinders. It is used in medicine for administration instead of air in certain cases, and finds its most important outlet in welding when it is mixed with hydrogen or acetylene gases, and this mixture burned to give very high temperatures. Liquid oxygen is commonly employed as a rocket propellant, being used to oxidise the combustible fuel.

NITROGEN

Like oxygen, nitrogen is diatomic, i.e., its molecule consists of two atoms, N_2. Its molecular weight is again twice the atomic weight, namely, $2 \times 14 = 28$. It is the major component of the air, but is also found in a combined form in certain minerals, and also in all living matter, chiefly in proteins.

AIR: ITS COMPONENTS 131

PREPARATION
Nitrogen is manufactured, as is oxygen, by the fractional distillation of liquid air.
On a laboratory scale there are many methods available. It may, for example, be made by progressively removing all the other major constituents from air. In this case air would be passed through the following:
(a) Sodium hydroxide solution—to remove carbon dioxide:
$$2NaOH + CO_2 = Na_2CO_3 + H_2O$$
(b) Alkaline pyrogallol solution *or* a tube containing heated copper: in either case the oxygen is removed; the equation for the latter is:
$$2Cu + O_2 = 2CuO$$

This gives nitrogen free from oxygen and carbon dioxide. It will contain water vapour and the inert gases. The former can be eliminated by drying at a suitable stage, but the latter cannot be removed by any chemical means. Fortunately the inert gases do not interfere in any of the tests likely to be made on the product. The gas is collected over water, or (if required dry) in a similar collecting apparatus but with water replaced by mercury.

A more convenient method does not start with the air. This uses the fact that a substance known as ammonium nitrite is very unstable and, on warming, breaks up into water and nitrogen. It is, in fact, difficult to prevent this compound from decomposing, and so it is not kept but is made, when required, by mixing solutions of sodium nitrite and ammonium chloride. The resultant solution is then warmed, the presence of a large volume of water preventing an explosive reaction. The reactions are:

$$NaNO_2 + NH_4Cl = NaCl + NH_4NO_2$$
$$NH_4NO_2 = N_2 + 2H_2O$$

The apparatus is shown in Fig. 15.3, with the gas collected over water.

PHYSICAL PROPERTIES
Nitrogen is without colour, taste or smell; it is less soluble in water than oxygen and has a lower boiling point.

CHEMICAL PROPERTIES
Nitrogen is chemically very unreactive. It will not support combustion except in the cases of a few very reactive metals; magnesium, for instance, burns in it with some difficulty on strong heating to

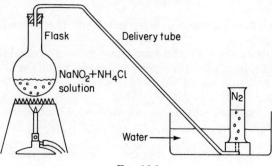

Fig. 15.3

give magnesium nitride:

$$3Mg + N_2 = Mg_3N_2$$

It is also non-inflammable in all other gases, though at very high temperatures it forms nitric oxide in small amounts with oxygen (e.g., in a lightning flash):

$$N_2 + O_2 = 2NO$$

The most useful reaction it has is its combination with hydrogen to form ammonia. This is the chief source of the latter gas. The conditions are not easy to reproduce in the laboratory, and involve pressures of several hundred atmospheres, temperatures around 500 K, as well as a catalyst containing several metal oxides:

$$N_2 + 3H_2 = 2NH_3$$

Apart from these reactions, and one or two others, nitrogen may be said to be nearly as inert as the argon family of 'inert gases'. It will be noted later (p. 149) that certain plants possessing the symbiotic bacteria can utilise nitrogen directly; this is all the more remarkable in view of the reluctance of the gas to react with most other things.

Uses: Nitrogen is chiefly used in the manufacture of ammonia (Haber process) and in providing an inert atmosphere in certain reaction-vessels, electrical equipment, etc. Liquid nitrogen is a refrigerant and has been used for freeze-drying baked goods.

CARBON DIOXIDE

This gas has the formula CO_2 and a molecular weight of 44. We have seen that it occurs in the atmosphere to a small but fairly

AIR: ITS COMPONENTS 133

constant extent, and later the significance of this will be discussed (p. 140).

PREPARATION

1. The burning of carbon and carbon compounds

Given sufficient oxygen or air, carbon and nearly all its compounds will burn to form carbon dioxide; restricted supplies of oxygen lead to the formation of carbon monoxide. Of course, with carbon compounds other substances are produced at the same time, e.g., water; obviously it is simpler to burn carbon itself. Industrially coke would be used; in the laboratory it would be most convenient to heat charcoal. Both of these are slightly impure forms of carbon:

$$C + O_2 = CO_2$$

An apparatus used for the preparation of carbon dioxide in the laboratory is given in Fig. 15.4. The gas can be collected over water in which it is not too soluble; it will probably be contaminated by some oxygen, as the supply of this must be plentiful to prevent carbon monoxide formation.

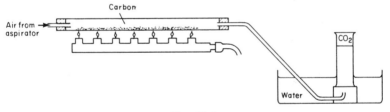

FIG. 15.4

2. Fermentation

This is a general term which means a chemical reaction brought about by micro-organisms; these secrete complex organic substances known as *enzymes* which are in fact complicated catalysts. Most fermentations are slow and produce gaseous products. Many reactions fall into this category. We are concerned only with those in which carbon dioxide results, and of these the most important by far is the conversion of glucose to alcohol and carbon dioxide under the influence of the enzyme zymase which is present in yeast.

To produce carbon dioxide by this means the most usual way is to allow sugar solutions to stand in contact with yeast. A preliminary reaction is the breakdown of sugar, or sucrose as it is often known, into two simpler molecules, fructose and glucose (*cf*. p. 209).

Both these are attacked by the zymase, and they themselves produced by the action of another enzyme present in yeast, invertase. As they both have the formula $C_6H_{12}O_6$, and as sucrose is $C_{12}H_{22}O_{11}$, the sequence of reactions can be written as follows, the name of the enzyme being indicated over the arrow:

$$C_{12}H_{22}O_{11} + H_2O \xrightarrow{invertase} C_6H_{12}O_6 + C_6H_{12}O_6$$
sucrose $\qquad\qquad\qquad\qquad$ fructose $\quad$ glucose

$$C_6H_{12}O_6 \xrightarrow{zymase} 2C_2H_5OH + 2CO_2$$
glucose or $\qquad\qquad$ ethanol
fructose $\qquad\qquad\quad$ ('alcohol')

This, of course, is the reaction used in the preparation of many alcoholic drinks (e.g. brandy), although others are formed from different starting materials. It also takes place on the addition of yeast to dough, when the gas produced causes the mixture to rise. Its employment in bakery practice has been considerably reduced by the advent of CBP.

As the substance from which the enzymes are derived, yeast, is a living organism, conditions have to be maintained in which it can live. This means the presence of small amounts of nutrients (inorganic salts as phosphates, usually present as impurities), and the absence of excesive heat. Yeast cells cannot survive over about 60°C, and therefore in baking fermentation occurs only in the early stages. The optimum temperature for working is between 25°C and 30°C. Since yeast is also killed by alcohol in more than 14% concentration, fermentation will automatically cease by the time that amount of alcohol has been formed (and probably a good while before).

As a laboratory method for making carbon dioxide this is fairly useless, mainly on account of the long time for appreciable quantities to be formed. If it is desired to show the reaction, however, the best way is to leave an inverted test-tube, filled with water, in a beaker containing sugar solution with a little yeast at the bottom, vertically beneath the test-tube opening. After some days, or possibly hours, the tube will be filled with the gas that has displaced the water from it.

3. *From carbonates and bicarbonates*

In Chapter 19 we shall see that carbon dioxide is very easily obtained from these salts of carbonic acid. We can summarise here the reactions that can be used:
  (a) Dilute acid on all carbonates and bicarbonates—the usual acids are hydrochloric or sulphuric, though nearly all others can be employed.

AIR: ITS COMPONENTS

(b) Heat on bicarbonates—the only bicarbonates available for this purpose are those of sodium and potassium.

(c) Heat on the carbonates of the 'heavy metals'—these include, broadly, all those except sodium and potassium.

For laboratory use we shall select two examples. First, we can heat sodium bicarbonate. The apparatus is identical with that in Fig. 15.2 except that the boiling-tube contains the bicarbonate instead of the potassium chlorate and manganese (IV) oxide. The carbon dioxide can be collected over water, or by displacement of air, as in the next example.

A second useful way of preparing carbon dioxide employs an apparatus often used to make gases when heat is not required but the action is of a liquid on a solid. This is the action of dilute hydrochloric acid on calcium carbonate. The main part of the apparatus (Fig. 15.5) is a Woulff's bottle, a heavy vessel made of thick glass which is easy to handle but cannot be heated. The gas could be collected over water as above, but use can be made of its greater density than air, which it will displace from the gas-jar.

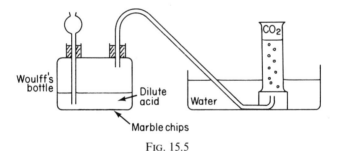

FIG. 15.5

The equations for these two reactions are

$$2NaHCO_3 = Na_2CO_3 + H_2O + CO_2$$
$$CaCO_3 + 2HCl = CaCl_2 + H_2O + CO_2$$

The production of carbon dioxide from baking-powders will be considered in Chapter 19, p. 183.

Physical Properties

Carbon dioxide is a colourless, odourless gas. It is about one and a half times as dense as air. It can be condensed to a solid far more easily than any other common gas, and this snow-like powder vaporises at $-78°C$. Although carbon dioxide is non-poisonous it cannot support animal life—hence the reputation of such places as the Death Valley where thick blankets of the gas cover the ground, having issued from crevices in the rocks. It is slightly soluble in water, more so under pressure.

Chemical Properties

Carbon dioxide neither burns nor supports combustion (except in unusual cases like magnesium which will continue to be oxidised if sufficiently hot). It can, however, be reduced to the monoxide if passed over heated carbon:

$$CO_2 + C = 2CO$$

When the gas dissolves in water, the solution has a slightly acidic reaction and turns blue litmus a claret colour. This happens because a weak acid (carbonic acid) has been formed. As this is unstable, attempts to isolate it by boiling of the water inevitably fail. All that occurs is dissociation back to water and carbon dioxide and one is left with a frustratingly clean and empty vessel. The formation of carbonic acid is thus reversible and can be written:

$$H_2O + CO_2 \rightleftharpoons H_2CO_3$$
$$\text{carbonic acid}$$

Since carbon dioxide forms an acid with water, and since acids react with alkalis, we could expect that alkalis would combine with the gas. This is what does in fact happen, and the resultant products are carbonates in the first place, and bicarbonates if passage of carbon dioxide is prolonged. With sodium hydroxide, one obtains sodium carbonate and then bicarbonate:

$$2NaOH + CO_2 = Na_2CO_3 + H_2O$$
$$Na_2CO_3 + H_2O + CO_2 = 2NaHCO_3$$

As both sodium salts are soluble, nothing is visible apart from the disappearance of the carbon dioxide bubbles. With calcium hydroxide, however, a different effect is seen for the first product (calcium carbonate) is insoluble, but the bicarbonate dissolves. What is seen, therefore, when carbon dioxide is passed into calcium hydroxide solution (or 'lime-water') is the appearance and subsequent disappearance of a precipitate. As few gases give this effect, this is the commonest test for carbon dioxide:

$$Ca(OH)_2 + CO_2 = CaCO_3\downarrow + H_2O$$
$$CaCO_3 + H_2O + CO_2 = Ca(HCO_3)_2$$

Finally, mention must be made of the complex but immensely important series of reactions by which carbon dioxide is converted in plants to carbohydrates in the process of photosynthesis.

AIR: ITS COMPONENTS 137

Uses: 1. Air exclusion: As carbon dioxide both fails to support combustion and forms a heavy blanket over a surface it is widely employed in fire-fighting. Many extinguishers contain an acid and a carbonate in separate compartments, and when operated (by depressing a plunger, etc.) they allow the reagents to come into contact, thus generating the gas; this is sometimes mixed with foam and directed through a nozzle on to the fire. Other makes use compressed carbon dioxide. In a different context carbon dioxide atmospheres extend the mould-free shelf-life of bakery products.

2. Refrigerant: Solid carbon dioxide ('dry ice', etc.) is a valuable refrigerant as it turns straight to gas when warmed. It has been used for such diverse purposes as keeping ice-cream cool and spraying on clouds to form artificial rain.

3. Medicine: Mixed with oxygen, small amounts of carbon dioxide stimulate respiration, and so are used to revive breathing after shock, etc.

4. Aerated waters: Use of carbon dioxide in solution under pressure as an addition to other drinks dates back to Priestley in the eighteenth century. 'Soda-waters' and other similar minerals on being poured out give up much of their gas as the pressure is now lower (atmospheric) and the solubility in water less. The carbon dioxide thus comes off as small bubbles.

INERT GASES

The remaining constituent of the air is in fact a mixture of five other gases, present in approximate proportions by volume as shown in Table 15.2.

TABLE 15.2

Gas	Proportion by volume (%)
Argon	0.93
Neon	0.001 8
Helium	0.000 5
Krypton	0.000 1
Xenon	0.000 01

The gases are all characterised by almost complete chemical inertness. It could be said that their failure to react springs from their inability to form compounds, and this in turn arises from a zero valency. However, recent discoveries of inert gas compounds have made us modify our extreme ideas on their 'inertness', although it remains true that they are extremely unreactive.

Obtained by fractionation of liquid air, they are used in electric discharge tubes, electric light bulbs, etc. Because of their inertness they allow filaments and other parts to reach very high temperatures without attacking them.

Revision 15.2

1. Give a carefully labelled diagram for the laboratory preparation of oxygen, nitrogen and carbon dioxide, one method for each gas.
2. How could you distinguish between four gas-jars, each containing a different constituent of the air?
3. Describe and explain the uses of the following: gas-jar, beehive shelf, pneumatic trough, Woulff's bottle.
4. How would you obtain a specimen of nitrogen from the air? How would this differ from a sample from decomposition of ammonium nitrite?

Chapter 16

Air: Its Reactions

In the previous chapter we have seen that air is a mixture of nitrogen, oxygen, the 'inert gases', carbon dioxide, and water vapour. Our present concern is the kind of effect that air has upon some of the most important chemical reactions that take place on the earth, and especially those reactions in which it actually plays a direct part.

The most obvious point is that it is essential for all common cases of combustion. When substances burn, the oxygen of the air is used in forming oxidised compounds, and these are dispersed into the atmosphere. If they are solids (as metallic oxides, etc.) they soon settle out and are returned to the earth taking their oxygen with them. Gases (as the oxides of nitrogen and sulphur) add their toll to atmospheric pollution, but except at certain localities their proportion in air is negligible and, in any case, they are slowly absorbed by certain minerals such as limestone; twenty years ago one had only to visit many industrial cities of the north to see the corrosive action they have on limestone buildings, and in this process they are removed from the air. These reactions, therefore, have the net effect of removing some of the oxygen from the atmosphere. How it is replaced will be seen shortly. There are two other gaseous products of combustion (and of respiration) that are far more important than these just mentioned. They are carbon dioxide and water vapour both of which are produced from organic matter amongst other sources.

Water vapour (or steam) results from chemical changes such as the burning of organic matter. This includes fuel oil and town gas, both of which contain hydrocarbons. The simplest of these, methane, burns according to the equation:

$$CH_4 + 2O_2 = CO_2 + 2H_2O$$

Others form the same products but in different proportions. Water is also formed when hydrogen gas burns, and this also is present in town gas and other gaseous fuels:

$$2H_2 + O_2 = 2H_2O$$

It results from certain decompositions of other kinds; for instance, when bicarbonates are heated water is formed, and this can be illustrated by baking powder:

$$2NaHCO_3 = Na_2CO_3 + H_2O + CO_2$$

Many other purely chemical reactions remove oxygen from the air and replace it by water vapour. There is, however, a 'safety device' which prevents the water from reaching high concentrations in the air and that is the quite low limit to the maximum humidity; above this amount, which varies with temperature, additional water is simply taken up as vapour but precipitated as mist or rain. Thus, we have again, as a net result of all these processes, just a removal of oxygen in small amounts. It is also true that evaporation of seas and rivers contributes far more to the small amounts of water vapour that are present than do the chemical reactions mentioned. The way in which the humidity may be reduced by rise in temperature will be familiar to all who have worked in a poorly air-conditioned, but hot, bakehouse (*cf*. p. 54).

The other very important product of combustion in air is the substance carbon dioxide. Unlike water vapour, which cannot exceed a quite small concentration of the air, there is no upper limit whatever to carbon dioxide, and the fact that it does not keep building up to ever larger proportions of the air needs to be explained, and we shall see that the answer lies in the concept of the 'carbon cycle'. In examining this we shall also find the chief method by which the oxygen concentration is restored.

THE CARBON CYCLE

The carbon cycle in nature is one of its most astonishing happenings. It is a cycle of events whereby the rates of removal of carbon dioxide are so delicately balanced by the rates of its replacement that the total concentration (except in very localised areas like the immediate neighbourhood of furnaces, etc.) remains almost exactly constant the world over—about 0.03%. We shall consider the various processes producing this gas, then those removing it and finally summarise our conclusions in a simple diagram.

PROCESSES INTRODUCING CARBON DIOXIDE INTO THE ATMOSPHERE
The following are the main sources of atmospheric carbon dioxide:

1. Combustion

This has already been mentioned. When any organic material, living or dead, animal or vegetable, is burned in sufficient air all its carbon ends up as carbon dioxide. This includes the burning of coal and petroleum products (motor and aviation fuel, 'paraffin', lamp oil, etc.), these being formed from very old deposits of terrestrial and marine life respectively. The combustion of wood and animal waste is a similar process. Carbon itself, chiefly as coke or charcoal, gives carbon dioxide, and so does the combustion of carbon monoxide present in such fuels as 'water gas':

$$C + O_2 = CO_2$$
$$2CO + O_2 = 2CO_2$$

Not all these processes are man-made, of course. Forest fires must have raged often before man appeared.

2. Respiration

The reason for the importance of oxygen for life is that it is converted into carbon dioxide in the vital process of respiration of animals, fishes and (under certain conditions) plants. It is from this reaction, or rather series of reactions, that the organism's energy is derived, for conversion of organic matter into carbon dioxide is a process that produces energy in some form. In the burning that takes place in a fire, this energy appears chiefly as heat. But in the body only part of it goes towards maintaining body temperature; some appears as muscular and nervous energy. In man and other animals, the oxygen is absorbed into the bloodstream in the lungs, where it is loosely combined with the red pigment of the blood (haemoglobin) to form oxy-haemoglobin. In this form it is transmitted via the arteries to various parts of the body where it may be needed. It then forms carbon dioxide in the tissues, and this is transported back to the lungs through the veins and is exhaled into the air. Oxygen is also required by fishes, but here it is dissolved in the water though still ultimately coming from the air. Plants also need it, especially in the dark. Again, carbon dioxide is the result.

3. Fermentation

This is a process of the greatest importance. Without giving a formal definition, we can say that it is a chemical change in which microorganisms have a powerful effect on the speed of action, and without which only negligible amounts of product would be obtained in a reasonable time. These microorganisms, or bacteria, secrete highly complicated substances known as enzymes, and it is these that are the catalysts. The most important case of fermentation is the decomposition of certain sugars of the formula $C_6H_{12}O_6$ into alcohol and carbon dioxide under the influence of enzymes present in yeast. The process is important in both brewing and bakery (see p. 209), but also occurs in the natural decomposition of fruit, etc. Glucose is fermented by the enzyme zymase:

$$C_6H_{12}O_6 = 2C_2H_6O + 2CO_2$$
(glucose)   (ethanol)

Related to this fermentation are the many bacterial changes known as putrefaction in which dead material decays into simple products, among them being carbon dioxide.

4. Mineral sources

Vast amounts of carbon exist in the earth as minerals (e.g., coal, petroleum, limestone, chalk, coral, etc., many, of course, ultimately organic in origin) and by various natural processes these sometimes give off carbon dioxide into the atmosphere. Volcanic action produces the gas, for instance. It is also found in high concentrations in certain parts of the world having issued from fissures in the ground and being unable to escape readily into the air as a whole. For example, at Death Gulch, California, the leakage occurs at the base of a deep valley where the gas tends to accumulate because of the lack of air currents. Of course, it does slowly diffuse into the upper atmosphere, but enough remains on the floor of the gulch to form a layer of several feet in depth; in this blanket of gas, life cannot continue, as little oxygen is present. Similar high concentrations of carbon dioxide are found elsewhere, e.g., in certain locations in Italy.

Other mineral sources of carbon dioxide include the man-made process of 'lime-burning', i.e., the manufacture of quicklime (calcium oxide) by heating limestone or chalk (calcium carbonate):

$$CaCO_3 = CaO + CO_2$$

Also, explosions in coal-mines give rise to carbon dioxide when 'fire-damp' (methane) or 'after-damp' (carbon monoxide, its partial combustion product) explode:

$$CH_4 + 2O_2 = CO_2 + 2H_2O$$

or,

$$2CH_4 + 3O_2 = 2CO + 4H_2O$$

then,

$$2CO + O_2 = 2CO_2$$

AIR: ITS REACTIONS 143

Thus by all these processes, and by others as well, carbon dioxide is continuously poured into the air. To explain why there is not a gradual 'build-up' of the gas we must take into account processes removing it from the air.

PROCESSES REMOVING CARBON DIOXIDE FROM THE ATMOSPHERE

1. Photosynthesis

This process is the absorption of carbon dioxide by plants and some marine organisms in the presence of sunlight. The carbon dioxide is converted ultimately into carbohydrates by reaction with water and under the catalytic influence of the green pigment chlorophyll. Sunlight is essential for this, and the reactions do not, therefore, occur in the dark; under those conditions plants absorb oxygen and evolve carbon dioxide, but when photosynthesis takes place oxygen is formed as well as the carbohydrate. Although the total process is enormously complicated in detail, it can be summarised by the simple equation:

$$xCO_2 + xH_2O = (CH_2O)_x + xO_2$$

The value of x will depend upon which carbohydrate is being considered. These substances are discussed on p. 207 of Chapter 22.

Photosynthesis is the process not only by which much of the atmospheric carbon dioxide is utilised, but also by which the oxygen taken up by the processes of combustion, respiration and the like is restored to the air. So far as plants are concerned, absorption takes place chiefly on the surface of green leaves. The marine organisms use dissolved carbon dioxide, but this has come from the atmosphere and the oxygen returns to it (apart from a small amount remaining in solution, and utilised in the respiration of other aquatic life).

2. Weathering of rocks

This is the only other large-scale process removing carbon dioxide from the air, and it does so to the order of 10^9 (a thousand million) tonnes per annum. A great deal of this action is due to dissolved carbon dioxide (when, as we have seen, carbonic acid is present) as rainwater, and then river water, runs over rocks composed of carbonates, chiefly of calcium and magnesium. The resultant water contains these metals as their bicarbonates and this causes part of the 'hardness' of some natural water (see p. 155):

$$CO_2 + H_2O = H_2CO_3$$
$$H_2CO_3 + CaCO_3 = Ca(HCO_3)_2$$
$$H_2CO_3 + MgCO_3 = Mg(HCO_3)_2$$

In this way, carbon is transferred both from rocks and the atmosphere into the sea. It will be restored to the air before this only if some artificial process is performed such as the boiling of the natural water, which simply reverses these three reactions.

SUMMARY OF THE CARBON CYCLE

In Fig. 16.1, the various processes affecting the amounts of carbon dioxide in the atmosphere are summarised. In addition, it shows the means whereby carbon is transferred between plants, animals, the sea and the earth. It is, of course, far from being a complete picture. Minor processes are not included, and we still do not know for certain where or how volcanic carbon dioxide originates, for instance, so it is not possible to include details of this. Our main emphasis here must be on those arrows leading to and from the atmospheric carbon dioxide.

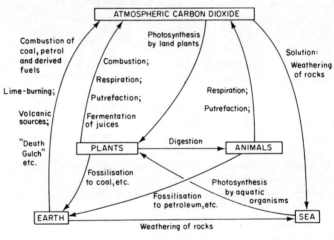

FIG. 16.1

THE NITROGEN CYCLE

The carbon cycle is of great importance to life because it plays a part in the vital processes both of plants and animals. All organic matter contains carbon—and this actually gives the modern definition of organic compounds, i.e., those containing carbon. All living tissues also contain the elements oxygen and hydrogen, these being readily available in water. But there is another element which is found in even the simplest forms of life, and which is just as essential. This is nitrogen, always found (with carbon, hydrogen, and oxygen) in proteins.

Nitrogen is made available to living organisms by a number of processes, and it is returned to its sources by other changes, all these operations constituting the nitrogen cycle. Now an important difference distinguishes this from the carbon cycle. In the latter, the atmosphere played a vital part, and the 'key' to the cycle was carbon dioxide. With the nitrogen cycle, however, the air is not so involved in the basic processes with which the cycle is concerned.

Broadly, the situation is this. Plants gain the nitrogen they require from the soil, where it is absorbed as nitrates. They then have their own means of converting the nitrates into proteins. When the plants die, and return to the soil, they liberate their nitrogen into it as ammonia. This is oxidised by the so-called 'nitrifying bacteria' to nitrates, which are then absorbed by more plants, etc. This simple picture is not altered much by inclusion of animal life; animals still derive their nitrogen from the soil ultimately, but by the method of eating plant materials. And when they die (or excrete) their nitrogen is also returned to the ground where it eventually forms ammonia.

This is the basic form of the nitrogen cycle, and it does not involve the atmosphere. It may be summarised as shown in Fig. 16.2 where X represents the action of the nitrifying bacteria (oxidation). In fact, however, matters are not quite as simple as this, and the cycle may be said not to be 100% efficient; in particular, leakages of nitrogen occur, causing certain amounts to be no longer available for plants and, therefore, animals also. This is obviously a serious thing, but, fortunately, other processes exist which help to redress the balance and give additional sources of nitrogen to supplement those of the main cycle. We shall consider both of these aspects at this point.

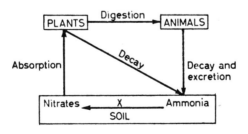

X = action of the nitrifying bacteria

FIG. 16.2

Factors Reducing the Amount of Available Nitrogen for Plants

1. Release of nitrogen to the atmosphere

Instead of all the ammonia in the soil being oxidised to nitrates at X, some becomes converted into gaseous nitrogen, N_2, and escapes into the air. Also, some of the nitrates that are formed are decomposed into nitrogen gas by the 'denitrifying bacteria'. The net result is that nitrogen that could have been utilised for living matter becomes nitrogen gas which is generally useless for life as it is so inert. It is these side-reactions that make the air a participant in the nitrogen cycle, and this is why we discuss the whole matter here.

2. Disposal of animal matter into the sea

This constitutes the other main source of nitrogen 'wastage'. Obviously nitrogen transferred to the sea is no longer available for land-plant life. This is largely a man-made difficulty, as modern methods of sewage disposal are responsible. Until comparatively recent times, human waste was returned directly to the ground. The well-known result of this lack of sanitation was, of course, the succession of waves of epidemics that swept Europe and elsewhere during the Middle Ages and since. But since the population density of towns has increased so much, the consequences of such primitive measures would now be so disastrous that new methods had to be developed.

As a result sewage disposal has become a skilled, if not a very attractive, occupation. After a variety of treatments, involving bacterial actions of various sorts, the sewage from inland communities is made innocuous and the nitrogen in it is returned to the soil (often as concentrated fertiliser which is sold as such). But the practice for communities near the coasts is often to pump the waste to a suitable distance out to sea, with the consequence that the valuable nitrogen is lost to the land.

Two less important sources of loss will be mentioned.

3. Fossilisation of plants to coal

This process was also a means whereby carbon was diverted in plants from ending up as carbon dioxide. Here it means that nitrogen is also 'locked up' in the form of complex compounds in coal, and is not capable of being absorbed by other plants.

4. Washing away of nitrates from the soil into the sea

Again we have met a parallel case in the weathering of rocks with

AIR: ITS REACTIONS   147

formation of bicarbonates. In this case a small amount of nitrates and other soluble materials is dissolved by rain water and washed into rivers and eventually the sea.

If this were the only side of the picture the consequences would be disastrous, for all the nitrogen in the soil would gradually diminish, and would end up in the sea (as soluble salts), the earth (as unavailable compounds in coal) or in the air (as nitrogen gas). Not merely would plants and animals be starved of proteins but the proportion of nitrogen in the atmosphere would increase and, therefore, the ratio of oxygen would decrease. But, as the carbon cycle is so delicately balanced that the amount of carbon dioxide remains constant within exceedingly fine limits, so also does the amount of nitrogen in the atmosphere. And other processes act in the opposite direction and help to restore the balance by making available new sources of nitrogen to plant life. We shall turn now to these.

FACTORS INCREASING THE AMOUNT OF AVAILABLE NITROGEN FOR PLANTS

1. Fixation of atmospheric nitrogen

This phrase means the conversion of atmospheric nitrogen into forms which can be assimilated by plants. Such a process is occurring in the air all the time and is a reaction which takes place during a flash of lightning. At the very high temperatures momentarily attained, small amounts of nitric oxide are produced by combination between the two chief constituents of air. As the gases cool down, further oxidation to nitrogen dioxide occurs and this combines with water to give nitric acid which is washed into the soil by rain:

$$N_2 + O_2 = 2NO$$
$$2NO + O_2 = 2NO_2$$
$$H_2O + 2NO_2 = HNO_2 + HNO_3$$

HNO_2 is nitrous acid which is later converted chiefly to nitric acid.

This process of nature (which produces a surprisingly large quantity of nitric acid over the whole world each day) has been imitated in a man-made operation in which air is passed through a large electric arc. As a large-scale source of nitric acid and fertilisers, it is now quite obsolete, owing to the enormous expenditure on current. But it is interesting to note, in passing, that it is still used on a small scale to make nitrogen dioxide for flour-bleaching (see also p. 198). However, there is a large-scale method for conversion of atmospheric nitrogen into useful forms, and that is the Haber process. In this, nitrogen from the air and hydrogen are passed over a catalyst containing iron at high temperatures and pressures. The product is ammonia, which can be oxidised to nitric acid or converted into one of its salts and added directly to the soil:

$$N_2 + 3H_2 = 2NH_3$$

Thus the comparatively useless nitrogen is converted into a valuable food for living organisms.

2. Utilisation of nitrogen directly by plants

It has been stressed that normally nitrogen is too unreactive to be of any use to an organism. There is an immensely important exception to this, however, and this constitutes one of the chief natural ways of restoring lost nitrogen to the soil. Certain plants known as the *Leguminosae*, and including peas, beans and clover, are able to absorb nitrogen directly from the air. They convert this to foodstuffs by the action of the so-called 'symbiotic bacteria'. Hence they can grow on poor soils, at least so far as available nitrogen is concerned. They are the second chief means for keeping the proportion of atmospheric nitrogen constant. Other nitrogen-fixing organisms exist but are so far less important.

3. Liberation from coal

Nitrogen which has been locked up for millions of years in coal is liberated as ammonia when coal is 'carbonised' to make coal-gas and coke (though today very little coal-gas is made as the primary product). The ammonia is washed out of the gas and the resulting 'ammoniacal liquor' is neutralised with sulphuric acid to make ammonium sulphate in which form it is sold as a fertiliser.

4. Animal consumption of marine life

Although nitrogen is lost to the sea, some of it is taken up by aquatic organisms. These in turn become the food of fishes and thus acquire the form of proteins that are edible by animals. Thus the nitrogen does become available to animal life, but not plant life directly. However, it is returned to the terrestrial nitrogen cycle.

In these ways the total amount of available nitrogen tends to remain constant in the soil as a whole. There are, of course, local deficiencies and local concentrations, and modern agriculture takes these into account. Thus the American deposits of Chile saltpetre and guano have for a long time been transported elsewhere as nutrients for poor soils. Since the eighteenth century farmers have known the value of the 'rotation of crops', where a field is sown on successive years with crops that demand different kinds of nutrient. This is the reason for growing clover in fields that have just had some other kind of crop, e.g. wheat. The latter removes nitrogen from the soil, but this is in some measure restored by the clover through the action of its symbiotic bacteria in converting to proteins the nitrogen of the air.

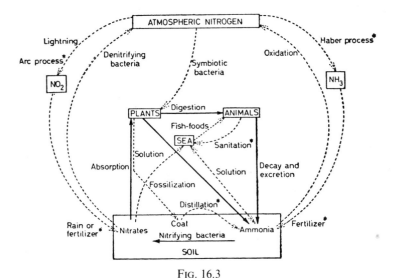

FIG. 16.3

Figure 16.3 is a fuller summary of the nitrogen cycle. The basic cycle is shown in unbroken lines, and the subsidiary changes in broken lines. Processes that are wholly or mainly man-made are indicated by an asterisk(*).

Chapter 17

Water

Water is the most abundant liquid on earth. It also has some very unusual properties, as will be seen later. These two facts combined focused attention on the composition of water from earliest times.

Whereas air turned out to be a mixture, water was found to be a compound. It consists of two elements, hydrogen (which means 'water-producer') and oxygen. These are in the ratio by weight 1:8, and with atomic weights of 1 and 16 respectively, they may be seen to be in a ratio of two atoms of hydrogen to one of oxygen. The molecular weight has been shown to be 18, so the formula of water becomes H_2O—the most famous chemical formula on record.

Oxygen has already been described. Hydrogen is a gas that we must now examine.

HYDROGEN

PREPARATION

Hydrogen can be made from water. One method uses electrolysis (p. 117), and the production of hydrogen and oxygen (only) in the ratio by volume 2:1 is strong evidence for the formula H_2O. Certain chemical substances will displace hydrogen from water when allowed to act upon it. They include those indicated in Table 17.1.

TABLE 17.1

Substance	Equation	Conditions
Sodium	$2Na + 2H_2O = 2NaOH + H_2$	Room temperature; vigorous.
Potassium	$2K + 2H_2O = 2KOH + H_2$	Room temperature; very vigorous.
Calcium	$Ca + 2H_2O = Ca(OH)_2 + H_2$	Room temperature; moderate rate.
Magnesium	$Mg + 2H_2O = Mg(OH)_2 + H_2$	Room temperature; very slow.
Magnesium	$Mg + H_2O = MgO + H_2$	Steam on hot metal; fast.
Carbon	$C + H_2O \rightleftharpoons CO + H_2$	Steam on hot coke; moderate.
Aluminium	$2Al + 6H_2O = 2Al(OH)_3 + 3H_2$	Appreciable if mercury present: room temperature.

WATER

On a commercial scale, the carbon method is [...] gas being called 'water gas'. Hydrogen is also [...] product in the petrochemicals industry, arising [...] processes to which petroleum fractions are subject [...] tory it is most convenient to use calcium, and the a [...] Fig. 17.1, where the test tube contains water th[...] displaced by the hydrogen that rises from the surface [...]

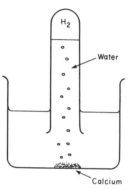

FIG. 17.1

Alternatively one can replace the calcium by lithium or calcium hydrides:

$$LiH + H_2O = LiOH + H_2$$
$$CaH_2 + 2H_2O = Ca(OH)_2 + 2H_2$$

Many metals do react with water to give hydrogen, but at rates that are so slow as to be almost imperceptible. The addition of acid, however, causes a great increase in rate of attack on the metal, and this gives a further method. The acid is usually dilute hydrochloric or sulphuric, but not nitric. Nearly all metals can be used except copper, silver, gold, mercury and some rare 'noble' metals like platinum. Some like lead are rather too slow in reacting. The following are just a few reactions, all occurring at room temperature, and capable of being carried out in an apparatus of the kind shown in Fig. 17.2; all acids are dilute:

$$Zn + H_2SO_4 = ZnSO_4 + H_2$$
$$Zn + 2HCl = ZnCl_2 + H_2$$
$$Mg + H_2SO_4 = MgSO_4 + H_2$$
$$Fe + H_2SO_4 = FeSO_4 + H_2$$
$$2Al + 6HCl = 2AlCl_3 + 3H_2$$

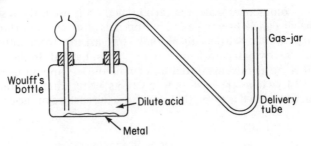

Fig. 17.2

Alternatively, the apparatus shown in Fig. 15.5 can be used, with the metal in place of the marble.

PHYSICAL PROPERTIES

Hydrogen is a colourless, odourless, tasteless gas, insoluble in water. It is not poisonous, but cannot support life. It has the distinction of being the lightest gas known; it has only 1/14th of the density of air, and so can be collected by the method of collection in Fig. 17.2.

CHEMICAL PROPERTIES

Although hydrogen will not support combustion, it will burn in air or oxygen with the greatest ease. Mixtures of these gases with hydrogen can be exploded by a spark:

$$2H_2 + O_2 = 2H_2O$$

If a hydrogen flame in air is allowed to impinge on a cool surface, drops of liquid are formed by condensation and may be identified as water. This is further evidence that water contains hydrogen and oxygen.

This reaction illustrates a fundamental property of hydrogen: its great affinity for oxygen. Not only will the gas combine with free oxygen (O_2); it will also remove oxygen from many substances in which oxygen is combined chemically. This process is *reduction* (see p. 188). Hydrogen will reduce many heated metal oxides, for example. Only the most stable, as those of aluminium and sodium, are unaffected; others as copper oxide are easily reduced:

$$CuO + H_2 = Cu + H_2O$$

Other elements that combine with hydrogen include chlorine, explosively in bright sunlight,

$$H_2 + Cl_2 = 2HCl$$

and nitrogen, in the presence of catalysts and under pressure,

$$N_2 + 3H_2 = 2NH_3$$

Uses: Because of its low density, hydrogen has been used in balloons and airships. It is chiefly employed today in welding, when a mixture with oxygen is burned to give local regions of very high temperatures, and in the manufacture of ammonia (NH_3) and modified natural fats as margarine. The latter is derived from vegetable oils which can take up hydrogen to form semi-solids, resembling butter, in the presence of a nickel catalyst. Hydrogen is not only a product of the petrochemicals industry; it is also used by that industry for various purposes, such as the removal of sulphur from petrol.

Revision 17.1

1. Describe two experiments to show that water contains hydrogen and oxygen.
2. Explain the 'dew' that appears on a kettle just after it has been placed on a gas-ring but rapidly disappears.
3. How could you distinguish hydrogen from oxygen and nitrogen?
4. Draw a diagram for the preparation of hydrogen from magnesium and dilute acid.

PHYSICAL PROPERTIES OF WATER

The most obvious things to say here are not quite true. Water is not, for instance, completely colourless, but in large quantities has a faint blue colour. Nor is it true that it is tasteless; the liquid we call 'drinking water' may appear to have no taste, but it is not pure water; water freed from most impurities by distillation does possess a peculiar 'flat' taste.

These are small points, however. At atmospheric pressure of 760 mm of mercury, water boils at 100°C and ice melts at 0°C, these being, of course, the fixed points of the Celsius scale. The density of water at 16°C is another convenient standard for other liquids and solids, being taken as unity.

Water has a number of exceptional properties, and these include the following:

1. Maximum density above its freezing point: When most things cool, they contract; their volume becomes smaller, and their density therefore increases. We should consequently expect that as water was cooled towards it freezing point its density would show a continuous increase until 0°C. In fact what does happen is that the density increases until 4°C, but with further cooling it then decreases till 0°C when freezing takes place. Thus we have the maximum density not at 0°C but at 4°C. Hence, when water cools it contracts till 4°C, but then expands until it freezes. If both ends of a pipe are blocked by ice for instance, and water in it cools from, say, 2°C to 0°C, it will try to expand and in so doing may well burst the pipe; this is one disadvantage of using a liquid such as water. On the other hand this expansive ability has been turned to good effect in the breaking-down of rocks over countless years to smaller units capable of being tilled as soil.

2. *High specific heat* (*cf.* p. 41): Water has the highest specific heat of all liquids. The simple consequence of this is that a large amount of heat energy has to be put in to a given weight of water to raise its temperature by a stated amount, and this means more fuel is needed than for other liquids. It also means that a longer time than for other substances is needed for water to heat up or cool down; hence variations in sea temperature are less than variations in air temperature over the same periods of time in the same localities. Further, materials rich in water (as jam) require fairly long times after removal from ovens before handling.

3. *Solvent ability:* Water has a unique ability as a solvent. It can dissolve countless substances of widely different types, sometimes in small quantities, it is true. Not merely does it dissolve them; it often permits them to ionise. It is for these reasons that most simple chemical reactions are studied in the presence of water. It also accounts for the state of many rocks on the earth's surface, much material having been at one stage in solution in the sea.

CHEMICAL PROPERTIES OF WATER

Water is rarely met with in a pure condition, and many of the reactions that we associate with it are not shown with pure water. Included among these is the corrosion of metals which is further considered in Chapter 20. Some metals are attacked by pure water, however, and a list of these is given in Table 17.1. Carbon is another element attacked by steam when it is brought to a red heat:

$$C + H_2O \rightleftharpoons CO + H_2$$

In the above reactions the hydrogen of water is replaced by another element. Sometimes the oxygen can be replaced, as when water with dissolved chlorine is exposed to light:

$$2Cl_2 + 2H_2O = 4HCl + O_2$$

WATER

The other main reactions of water take place when it combines with other oxides. If a metal oxide dissolves in water (and not many do) it forms an alkaline solution, the alkali being the metal hydroxide; the only important examples are:

$$Na_2O + H_2O = 2NaOH$$
$$K_2O + H_2O = 2KOH$$
$$CaO + H_2O = Ca(OH)_2$$

A similar alkali is formed from ammonia:

$$NH_3 + H_2O = NH_4OH$$

If an oxide of a non-metal dissolves and reacts it forms an acid, as the following:

$$CO_2 + H_2O = H_2CO_3 \quad \text{(carbonic)}$$
$$SO_2 + H_2O = H_2SO_3 \quad \text{(sulphurous)}$$
$$SO_3 + H_2O = H_2SO_4 \quad \text{(sulphuric)}$$
$$P_2O_5 + H_2O = 2HPO_3 \quad \text{(phosphoric)}$$
$$N_2O_4 + 2H_2O = HNO_2 + HNO_3 \quad \text{(nitrous and nitric)}$$

Water undergoes a number of other miscellaneous reactions of which the most familiar is probably its action on calcium carbide to give the inflammable gas acetylene, formerly used for lighting and now an important welding gas and chemical intermediate:

$$CaC_2 + 2H_2O = Ca(OH)_2 + C_2H_2$$

Revision 17.2

1. Give two reactions in each case whereby hydrogen and oxygen may be obtained from water.
2. Write equations for the action of water on carbon dioxide, sulphur trioxide, calcium carbide and calcium oxide.
3. Show how the unusual physical properties of water have had desirable effects in nature.

HARDNESS OF WATER

This rather strange term refers to a characteristic of certain kinds of tap water. In some areas, the water from the taps fails to give easily a lather with soap, instead tending to form a scum on the surface. This applies only to soaps and soap powders, not to synthetic detergents. Water that behaves in this way is said to be 'hard', while that which lathers easily with soap is termed 'soft'.

We shall look first at the causes of this, then at some other results, and finally at the steps which may be taken to deal with the situation.

CAUSES OF HARDNESS IN WATER

The behaviour of hard water is attributable to impurities in the water. In particular, the presence of dissolved minerals is responsible. To see how they get there we must trace the tap water back to its source.

Ultimately it originates as rain, which is water in a fairly pure state. This falls on to rocks, soil, etc., and soon finds its way to lower levels, eventually entering streams, rivers or the sea. Before it reaches them, however, it will have passed over a number of rocks containing soluble compounds, especially the chlorides and sulphates of calcium and magnesium—$CaCl_2$, $CaSO_4$, $MgCl_2$ and $MgSO_4$. These tend to dissolve and so are found in appreciable amounts in the water taken by the water boards for domestic use. If the only treatment at the water-works is filtration (to remove suspended matter) and sterilisation (to kill germs), these salts will be present in tap water.

But this is not all. Rain water, despite its purity, does contain some dissolved carbon dioxide which it has taken up from the atmosphere. As we have seen (p. 143) a solution of this kind will turn carbonates into bicarbonates. As all the latter are soluble, passage of this solution over insoluble carbonates will tend to dissolve out some of the solid by converting it into soluble bicarbonate. This is just what occurs when rain water passes over limestone, chalk, and other minerals that are largely calcium carbonate, and similarly with magnesium carbonate deposits:

$$CaCO_3 + H_2O + CO_2 = Ca(HCO_3)_2$$
$$MgCO_3 + H_2O + CO_2 = Mg(HCO_3)_2$$

Thus tap water may contain varying amounts of at least six salts: chlorides, sulphates and bicarbonates of calcium and magnesium. Why should it affect the lathering of soap?

Soap is another salt (or a mixture of several) derived from organic acids in fat, and sodium. Let us consider the chief component, sodium stearate, which we shall write as NaSt, the symbol 'St' standing for an organic radical called the stearate group. Now although soap is (sparingly) soluble, the stearates of calcium and magnesium are not. Hence, when soap is added to hard water, some of it reacts with the calcium and magnesium salts to form these insoluble compounds which form a scum on the surface. Not until all the calcium and magnesium have ended up as stearates is the soap available for its chief function—to form a lather and remove dirt. For the calcium compounds the equations are:

$$CaCl_2 + 2\,Na\overline{St} = Ca\overline{St}_2 \downarrow + 2NaCl$$
$$CaSO_4 + 2\,Na\overline{St} = Ca\overline{St}_2 \downarrow + Na_2SO_4$$
$$Ca(HCO_3)_2 + 2\,Na\overline{St} = Ca\overline{St}_2 \downarrow + 2NaHCO_3$$

The equations for magnesium compounds are similar.

OTHER EFFECTS OF HARDNESS IN WATER

Originally, the chief disadvantage of hard water, the wastefulness of soap, is now no longer so important, chiefly on account of the synthetic detergents; some of these behave equally well in hard and soft water. Today the most serious drawback of hard water is the 'furring' of hot water pipes. If the hardness is due to bicarbonates, the water will deposit the carbonates on heating, for bicarbonates cannot exist in hot water:

$$Ca(HCO_3)_2 = CaCO_3 \downarrow + H_2O + CO_2$$
$$Mg(HCO_3)_2 = MgCO_3 \downarrow + H_2O + CO_2$$

This accounts for the 'fur' on the bottom of kettles (usually darkened by traces of iron also present). If, however, the reaction takes place in pipes, there will accumulate a deposit of carbonate on the inside and this will continue until they are blocked up completely. In the extreme case an explosion may result, but even with only thin layers of deposit the efficiency of a heater is much reduced because these layers are insulators and so more heat has to be added to boil the water inside.

In the bakery processes hard water has little disadvantage. Because it is slightly alkaline (if containing bicarbonates) it may cause a small reduction in the rate of fermentation of a dough, but this is too slight to matter greatly. Also, the dough may be slightly tighter.

Hard water has one advantage over soft. Regular intake of calcium compounds is essential to the well-being and growth of bones and teeth, especially in children. Drinking of very soft water only may, therefore, cause calcium deficiency, with subsequent danger of diseases like rickets and dental decay. It has also been associated with a prevalence of certain forms of heart disease.

REMOVAL OF HARDNESS IN WATER

Many processes for 'softening' hard water have been suggested, and the following are the most important:

1. Distillation: This removes all traces of mineral salts, but is prohibitively expensive (except on a limited laboratory scale) owing to the large amount of heat required to boil the water (*cf.* p. 154).

2. *Heating* (not necessarily to boiling-point): this destroys the bicarbonates, as in the last equations. The carbonates are precipitated, and thereafter have no action on the soap, being out of solution. The alkalinity due to them also vanishes, and of course there is now no possibility of this reaction occurring in hot pipes. As heating has no effect on either sulphates or chlorides, this treatment removes only part of the hardness—that due to bicarbonates. For this reason we distinguish two kinds of hardness, defined as below:

Temporary hardness—the hardness in water that is caused by the presence of dissolved bicarbonates, and which can be removed by heating.

Permanent hardness—the hardness resulting from the presence of dissolved chlorides and sulphates in water, and being incapable of being destroyed by heating.

3. *Addition of soda (sodium carbonate) solution:* This precipitates all the calcium and magnesium as carbonates, irrespective of the form they were in originally, e.g.,

$$CaSO_4 + Na_2CO_3 = CaCO_3 \downarrow + Na_2SO_4$$
$$MgCl_2 + Na_2CO_3 = MgCO_3 \downarrow + 2NaCl$$
$$Ca(HCO_3)_2 + Na_2CO_3 = CaCO_3 \downarrow + 2NaHCO_3$$

The precipitated carbonates and the dissolved sodium salts have no further effects. This process is sometimes used industrially.

4. *Addition of lime (calcium hydroxide):* Here we have another method (Clark's process) for removal of temporary hardness only. The lime is added in calculated quantities so that it just neutralises the bicarbonates:

$$Ca(HCO_3)_2 + Ca(OH)_2 = 2CaCO_3 \downarrow + 2H_2O$$
$$Mg(HCO_3)_2 + Ca(OH)_2 = CaCO_3 \downarrow + MgCO_3 \downarrow + 2H_2O$$

It is important that too much lime is not added or this will add to the hardness.

5. *Method of ion exchange:* This method is used in domestic, laboratory and industrial installations. Certain synthetic resins are used which contain sodium ions which can be 'stripped off' and replaced by calcium or magnesium ions (and others) when a solution containing these ions percolates over them. Thus the salts of these metals are changed into the corresponding sodium salts which do not cause hardness. Similarly the resins become poorer in sodium and richer in the other metals; they can be regenerated by pouring a concentrated solution of salt over them when the 'force of numbers' of sodium ions causes them to take the place of the others on the resin. This brings matters back to where they were.

WATER 159

The resins are generally of the type known as sodium aluminium silicates, or commercially as 'permutits'. This method uses up only salt, which is extremely cheap, and has the additional advantage of dealing with both types of hardness.

6. *Use of sodium hexametaphosphate* $(NaPO_3)_6$: This compound, on addition to water containing calcium and magnesium ions, converts them to stable, soluble complex compounds which have no action on soap solutions and are not decomposed by heat. This was used extensively by railway authorities for locomotive boilers.

Revision 17.3
1. Write equations for the action of soap on three magnesium salts likely to be present in hard water.
2. Give two methods for removal of temporary hardness only.
3. Define the terms 'hard water', 'temporary hardness' and 'permanent hardness'.
4. What do you think are the chief disadvantages of hard water for use in a bakery?
5. Given a supply of water known to contain calcium bicarbonate, suggest how you could obtain from it samples of calcium carbonate and carbon dioxide gas.

Chapter 18

Acids, Bases and Salts—I

NATURE OF ACIDS, BASES AND SALTS

Most people know that pickling cabbage is turned a reddish colour by vinegar; they probably do not know also that if one adds lime to this, it will return to its original bluish colour, while more vinegar reddens it again, and so on.

This behaviour of cabbage leaves can be paralleled in the laboratory rather more neatly by using the vegetable extract known as litmus; a similar colour change is observed.

Vinegar is far from being the only substance to have this effect, and many other compounds than lime can counteract it. Accordingly, we can group together into two classes all those substances that have any effect on litmus. Those that turn it red we shall call acids and those that turn it blue shall be alkalis. These are not precise definitions—we shall come to those later.

The group we have termed acids have in common certain other features:

(i) Have a sour taste.
(ii) Are often corrosive on the skin.
(iii) Liberate carbon dioxide from carbonates.
(iv) Give hydrogen with metals as iron and zinc.
(v) Have characteristic action on other materials termed 'indicators' which they turn definite colours. Apart from litmus, we have methyl orange (yellow to red), phenolphthalein (pink to colourless), and many others.

Unfortunately, most of these features have exceptions. Some acids (termed 'weak' acids) such as carbolic acid, do not give carbon dioxide with carbonates and are often without a sour taste. A few, such as nitric acid, give rise to hydrogen with metals only rarely. For the moment, therefore, we shall regard as acids only those compounds that have the above action on indicators.

Another group of substances has the opposite action on indicators and these are termed alkalis. These may have an additional

ACIDS, BASES AND SALTS—I

common feature in that their solutions have a soapy feel (actually because they attack the skin and form small amounts of soap from it). Alkalis are thus in one sense the opposite of acids. In fact they will combine with each other to form products that have no action on indicators, and these are called 'salts'. An alkali will neutralise an acid (and vice versa).

The following are examples of some common acids and alkalis:

Acids		Alkalis	
HCl	hydrochloric	NaOH	sodium hydroxide
H_2SO_4	sulphuric	KOH	potassium hydroxide
HNO_3	nitric	$Ca(OH)_2$	calcium hydroxide
H_2CO_3	carbonic	NH_4OH	ammonium hydroxide
H_6C_6O	carbolic		
HCN	prussic		
$H_4C_2O_2$	acetic		
H_2SO_3	sulphurous		
HNO_2	nitrous		

Examination of the above lists shows clearly that all acids mentioned have hydrogen as one constituent, while all alkalis have the group of atoms OH known as the hydroxide radical. When acids and alkalis combine, the hydrogen of the acid is replaced by the metal atom(s) from the alkali, while its hydrogen unites with the OH of the latter to form water. Thus we have:

$$\text{acid} + \text{alkali} = \text{salt} + \text{water}$$

It does not follow that all the hydrogen of the acid will be replaced. With sulphuric acid, for example, there are two series of salts with one or two atoms of hydrogen replaced. But with carbolic acid, only one of its six hydrogen atoms is ever replaceable in this way. Some examples of these reactions follow:

$$\begin{aligned}
HCl + NaOH &= NaCl + H_2O \\
H_2SO_4 + 2KOH &= K_2SO_4 + 2H_2O \\
2HNO_3 + Ca(OH)_2 &= Ca(NO_3)_2 + 2H_2O \\
H_2SO_4 + NaOH &= NaHSO_4 + H_2O
\end{aligned}$$

The number of replaceable hydrogen atoms that an acid molecule possesses is termed its basicity; hydrochloric acid is monobasic, sulphuric acid dibasic, and so on. With polybasic acids (more than one replaceable hydrogen atom) more than one series of salts can be formed, the nature of the product depending on the ratio of acid to alkali.

We can now go further and say that salts may be formed by many other methods than neutralisation of an acid by an alkali. We have already met the formation of carbonates by a precipitation reaction, for instance.

It is now possible to define salts without reference to their method of preparation; they are compounds derived from acids by total or partial replacement of hydrogen by a metal. Compounds of this type have much in common; all are solids, many soluble in water, and all with high melting points. They include compounds not derived from the four common alkalis; there are salts of copper, zinc, iron, magnesium, and of every other metal known. Consider copper sulphate, $CuSO_4$. This cannot be made from an alkali because copper hydroxide is not soluble and so has no action on indicators. But it can be made from sulphuric acid and copper oxide or hydroxide although neither is an alkali:

$$Cu(OH)_2 + H_2SO_4 = CuSO_4 + 2H_2O$$
$$CuO + H_2SO_4 = CuSO_4 + H_2O$$

Compounds like these that react with acids to give salts are termed bases, and only if they are water-soluble, and therefore have an action on indicators, are they alkalis as well. Hence

$$acid + base = salt + water$$

In the light of these ideas we shall now attempt rather more precise definitions of the terms we have been using:

Acids—compounds with one or more hydrogen atoms replaceable by a metal.

Bases—compounds capable of reacting with an acid so that the hydrogen of the latter is replaced by a metal (or ammonium).

Salts—the products formed along with water when an acid reacts with a base.

Alkalis—soluble bases.

Basicity of an acid—the maximum number of hydrogen atoms in the molecule that may be replaced by a metal.

Indicators—substances that change colour with acids and alkalis.

IONISATION THEORY OF ACIDS AND BASES

Acids, bases and salts have one thing in common: if they are soluble in water their solutions are always ionised (*cf.* p. 116). Let us take the substances hydrochloric acid, sodium hydroxide, and sodium chloride:

$$HCl \rightleftharpoons H^+ + Cl^-$$
$$NaOH \rightleftharpoons Na^+ + OH^-$$
$$NaCl \rightleftharpoons Na^+ + Cl^-$$

Water, on the other hand, is scarcely ionised at all. Hence, we may rewrite the reaction

$$NaOH + HCl = NaCl + H_2O$$

in terms of the ions involved:

$$Na^+ + OH^- + H^+ + Cl^- = Na^+ + Cl^- + H_2O$$

As both Na^+ and Cl^- appear on both sides of the equation we may cancel them out, and all that is left is:

$$OH^- + H^+ = H_2O$$

The OH^- comes from the alkali and the H^+ from the acid. This is really all that takes place in a neutralisation reaction; the rest of the reacting species do not matter in the sense that they are unchanged throughout.

It is true that salts may be regarded as being completely ionised in solution, but this is not necessarily so for acids or alkalis. That is why all indicators do not behave in the same way to all acids and alkalis. A weak acid has a relatively small degree of ionisation, and therefore a small concentration of hydrogen ions, responsible for its action on litmus. That is why sulphuric acid turns it bright red, carbonic acid only a claret colour, and prussic acid does not affect it. That is also why it is wrong to define an acid as something which 'turns blue litmus red'. Alkalis, similarly, are strong only if they are sufficiently ionised to give a high concentration of hydroxide ions.

We have said that water is scarcely ionised at all. If we try to make this more precise we can say that, at room temperature, the concentration of hydrogen ions in pure water is 1 ion for every 10 000 000 molecules, that is $1/10^7$, or 10^{-7}. Since water can only ionise as:

$$H_2O \rightleftharpoons H^+ + OH^-$$

the concentration of hydrogen ions is the same as that of hydroxide ions, 10^{-7}.

For a strong acid, however, the concentration of hydrogen ions may be very much higher, perhaps 1 in 10, or 10^{-1}. With an alkali the concentration of hydrogen ions can never be more than 10^{-7}, and may be as low as 10^{-14}.

To express the 'acidity' of a solution, a very useful scale has been devised known as the pH scale, defining pH as:

The pH of a solution is the negative of the logarithm (to base 10) of the hydrogen ion concentration.

A logarithmic scale is used simply as a matter of convenience to avoid cumbersome fractions.

A very acidic solution will have a hydrogen ion concentration of perhaps 10^{-1}, a very alkaline one of possibly 10^{-14}, while pure water has a value of 10^{-7}. Let us see what will be the pH values of these solutions:

Hydrogen ion concentration = 10^{-1} ; $\log_{10} = -\ 1.0$; pH = $\ 1.0$
Hydrogen ion concentration = 10^{-7} ; $\log_{10} = -\ 7.0$; pH = $\ 7.0$
Hydrogen ion concentration = 10^{-14}; $\log_{10} = -14.0$; pH = 14.0

Thus, in practice, we have a series of possible pH values ranging from 1 for very acidic solutions, through 7 for neutrality, to 14 for highly alkaline solutions. A solution of pH = 8.2, for instance, is mildly alkaline, while one of 3.4 is quite strongly acidic.

It is possible to find the pH of a solution by means of indicators, for each has a definite pH range over which it will change colour, and by using suitable combinations of indicators one can determine fairly accurately the pH of a given solution. 'Universal Indicator', which changes colour continuously over the whole pH range, is a complicated blend of many indicators. Indicated below are ranges of a number of indicators, with their colours at each end:

pH 1 2 3 4 5 6 7 8 9 10 11 12 13 14

PHENOLPHTHALEIN
colourless ⟵⟶ pink

LITMUS
red ⟵⟶ blue

METHYL RED
red ⟵⟶ yellow

METHYL ORANGE
red ⟵⟶ yellow

Revision 18.1
1. Enumerate the chief properties of acids.
2. Define acids, bases and salts.
3. Calculate the pH of solutions with hydrogen ion concentrations of 10^{-5}; 1.8×10^{-5}; 3.7×10^{-11}.

TITRATIONS

These form the most obvious part of 'volumetric analysis' which is the method for estimation of amounts of substances in solution by measurement of volumes taking part in reactions.

ACIDS, BASES AND SALTS—I

To understand the process we must introduce the concept of a molar solution.

Molar Solution—A solution containing one mole (molecular weight in grams) of solute per cubic decimetre (litre) of solution.

The molarity of a solution is the number of times that the solution is more concentrated than one that is molar (M). Thus a 2M solution has 2 moles per cubic decimetre, while a 0.1M solution (sometimes called decimolar) has only 1/10 of a mole per cubic decimetre.

It is important to specify exactly the molecular species being considered. Thus consider a solution of sodium hydroxide of concentration 40 g/dm³. This will be ionised according to the equation:

$$NaOH = Na^+ + OH^-$$

Since the molecular weight for sodium hydroxide is 40 (23 + 16 + 1), the solution is clearly molar with respect to NaOH. Since each 'molecule' of NaOH gives rise to one ion each of Na^+ and OH^- it is also molar with respect to these. Now suppose we take the case of magnesium chloride which ionises as:

$$MgCl_2 = Mg^{++} + 2Cl^-$$

A solution that is molar with respect to magnesium chloride is molar for magnesium ions but 2-molar for chloride ions.

This idea is easily applied to volumetric analysis. Consider the reaction:

$$NaOH + HCl = NaCl + H_2O$$

If the solutions exactly neutralise one another there must be the same number of moles of each. Let us suppose that sodium hydroxide is x-molar and its volume is a cm³. Then:

	1	dm³	of	M	NaOH	contains	1	mole
	1	dm³	of	xM	NaOH	contains	x	mole
	1	cm³	of	xM	NaOH	contains	$10^{-3} x$	mole
so	a	cm³	of	xM	NaOH	contains	$10^{-3} ax$	mole

If this exactly neutralises b cm³ of yM HCl, the number of moles here must be $10^{-3} by$. But as we have seen these must be equal, hence:

$$10^{-3}ax = 10^{-3}by,$$

or

$$ax = by$$

If we know the molarity of one solution (say, x) and can measure the volumes a and b experimentally, then y can be immediately calculated. We do of course need to know the reactions taking place. If for example 1 mole of one reagent reacts with more than one of another we need to take that into account in our calculations. Thus in the following reaction if b cm^3 of yM sodium hydroxide are involved in neutralisation then the number of moles of sulphuric acid will be half that required for HCl:

$$2NaOH + H_2SO_4 = NaSO_4 + 2H_2O$$

consequently,

$$2ax = by$$

To carry out a titration, we use a pipette to introduce an accurately known volume of one solution into a conical flask (often 25.0 cm^3). We then place the other solution in the burette, and allow it to run slowly into the flask, having noted the level it was at beforehand. In the flask we have placed some suitable indicator capable of registering when the reagent there has been just neutralised. Solution is added from the burette until the indicator shows that we have added just the right amount (the 'end-point'). The liquid level in the burette is noted again and the difference between the two readings is the total volume added. We now know both volumes (a and b), so if the molarity of either solution is known that of the other may be calculated from the appropriate equation.

It is often necessary to prepare a standard solution (one whose molarity is accurately known). For this purpose a known weight of solute is dissolved in solvent, and the solution is made up to a known volume in a standard flask, often 250 cm^3. A graduation mark on the neck indicates the level to which the liquid should rise. These items are illustrated in Fig. 18.1, which also shows the correct way to read the volume of a liquid in graduated vessels; the bottom of the meniscus must be on the line when viewed at eye-level.

We may be concerned to use the method of titration to estimate the amounts of acid or alkali in various materials. To do this, we should first dissolve a known weight of sample in water, and make up the volume to, say, 250 cm^3; portions, e.g., 25 cm^3, of this solution will then be pipetted into a conical flask and 2–3 drops of an indicator added. The standard acid (or alkali) is then added until the indicator just changes colour and the volume added is recorded.

Not all the solutions used in titration can be prepared directly to have exactly known concentration. For example, sodium hydroxide is so deliquescent that it is never possible that what is weighed out is pure, and contains no water. Compounds that can be weighed out and made up at once to give a standard solution are known as primary standards:

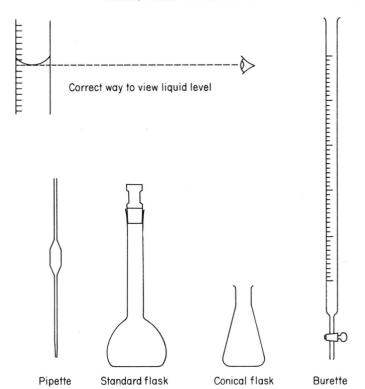

Fig. 18.1

Primary standards—substances from which standard solutions can be prepared from known weights in known volumes.

With other standard solutions, a preliminary titration against a solution of a primary standard has to be performed to establish their own concentration.

For estimating acids, it is often desirable to use instead of sodium and potassium hydroxide, carbonates of these metals or borax. Although not capable of affecting all indicators, and not being strong alkalis, they are much better than the caustic alkalis and can be used as primary standards.

In Table 18.1 are given some suitable combinations of acids and alkalis in volumetric analysis and Table 18.2 gives molecular weights for these compounds.

TABLE 18.1

Substances to be estimated	Suitable standards	Suitable indicators
HCl, H_2SO_4 or HNO_3	NaOH	any
	KOH	any
	†Na_2CO_3	methyl orange
	†K_2CO_3	methyl orange
	†$Na_2B_4O_7.10H_2O$	methyl orange
Organic acids	NaOH	phenolphthalein
	KOH	phenolphthalein
NaOH or KOH	HCl	any
	H_2SO_4	any
	HNO_3	any
	$H_4C_2O_2$	phenolphthalein
	†$H_2C_2O_4.2H_2O$	phenolphthalein
Na_2CO_3, K_2CO_3,	HCl	methyl orange
$KHCO_3$, or	H_2SO_4	methyl orange
NH_4OH	HNO_3	methyl orange

† primary standard

TABLE 18.2

Name	Formula	Molecular weight
Hydrochloric acid	HCl	36.5
Sulphuric acid	H_2SO_4	98
Nitric acid	HNO_3	63
Acetic acid	$H_4C_2O_2$	60
Oxalic acid	$H_2C_2O_4.2H_2O$	126
Sodium hydroxide	NaOH	40
Potassium hydroxide	KOH	56
Sodium carbonate	Na_2CO_3	106
Potassium carbonate	K_2CO_3	138
Borax	$Na_2B_4O_7.10H_2O$	381.4
Sodium bicarbonate	$NaHCO_3$	84
Potassium bicarbonate	$KHCO_3$	100
Ammonium hydroxide	NH_4OH	35

To illustrate the points so far, we shall now consider several examples.

ACIDS, BASES AND SALTS—I

Example 1

24.3 cm³ of 0.1M *hydrochloric acid was found by titration to be equivalent to 25.0 cm³ of a solution of sodium hydroxide. What was the molarity of the latter, and its concentration in* g/dm^3?

$$cm^3 \times \text{molarity of HCl} = cm^3 \times \text{molarity of NaOH}$$
$$\text{molarity of NaOH} = \frac{cm^3 \times \text{molarity of HCl}}{cm^3 \text{ of NaOH}}$$
$$= \frac{24.3 \times 0.1}{25.0}$$
$$= 0.097$$

Now, a molar solution of NaOH contains $40 g/dm^3$. Concentration of given solution is therefore:

$$0.097 \times 40 = 3.88 \text{ g}/dm^3$$

Example 2

A solution was prepared containing 1.30 g of pure sodium carbonate in 250 cm³. In titration against sulphuric acid, 28.6 cm³ of the acid were required to neutralise 25.0 cm³ of the sodium carbonate solution. Find the concentration of the acid as molarity and as g/dm^3.

A molar solution of anhydrous sodium carbonate contains $106 g/dm^3$, and therefore ($106 \times \frac{1}{4}$), or 26.50 g per 250 cm³.

$$\text{Molarity of sodium carbonate} = \frac{1.300}{26.50} = 0.049\ 05$$

Since in this reaction

$$Na_2CO_3 + H_2SO_4 = Na_2SO_4 + H_2O + CO_2$$

the molar ratio is 1:1,

$$cm^3 \times \text{molarity of } H_2SO_4 = cm^3 \times \text{molarity of } Na_2CO_3$$
$$\text{molarity of } H_2SO_4 = \frac{cm^3 \times \text{molarity of } Na_2CO_3}{cm^3 \text{ of } H_2SO_4}$$
$$= \frac{25.0 \times 0.049\ 05}{28.6}$$
$$= 0.042\ 85$$

Now, a molar solution of H_2SO_4 contains 98 g/dm^3. Concentration of given solution is therefore

$$0.042\ 85 \times 98 = 4.20 \text{ g}/dm^3$$

Example 3

To determine the amount of free alkali in a sample of soap, 1.23 g of soap were dissolved in water (containing some alcohol to facilitate solution), and the volume made up to 250 cm^3. In the presence of phenolphthalein, 25.0 cm^3 portions of this solution were titrated against 0.051M oxalic acid, of which 7.1 cm^3 were required for neutralisation. Calculate the percentage of free alkali in the soap, assuming it is all present as NaOH. Oxalic acid is dibasic.

cm^3 × molarity of NaOH = 2 × cm^3 × molarity of oxalic acid

$$\text{molarity of NaOH} = \frac{2 \times \text{cm}^3 \times \text{molarity of oxalic acid}}{\text{cm}^3 \text{ of NaOH}}$$

$$= \frac{2 \times 7.1 \times 0.051}{25.0}$$

$$= 0.029$$

Now, a molar solution of NaOH contains 40 g/dm^3. Concentration of NaOH in the soap solution is therefore:

$$0.029 \times 40 = 1.16 \text{ g/dm}^3$$
$$= 0.29 \text{ g per } 250 \text{ cm}^3$$

But weight of soap per 250 cm^3 was 1.23 g

$$\text{so percentage of NaOH in soap} = \frac{0.29}{1.23} \times 100$$
$$= 23.6\%$$

Example 4

To determine the percentage purity of sodium bicarbonate to be used in baking powder, 2.23 g were dissolved in water, and the solution made up to 250 cm^3, and 25.0 cm^3 portions were titrated against 0.105 molar hydrochloric acid. For neutralisation, 23.9 cm^3 of the acid were required. Calculate the percentage purity of the sample. The equation is:

$$\text{NaHCO}_3 + \text{HCl} = \text{NaCl} + \text{H}_2\text{O} + \text{CO}_2$$

cm^3 × molarity of NaHCO$_3$ = cm^3 × molarity of HCl;

$$\text{molarity of NaHCO}_3 = \frac{\text{cm}^3 \times \text{molarity of HCl}}{\text{cm}^3 \text{ of NaHCO}_3}$$

$$= \frac{23.9 \times 0.15}{25.0}$$

$$= 0.1004$$

Now, a molar solution of NaHCO$_3$ contains 84 g/dm^3. Concentration of given solution is therefore:

$$0.1004 \times 84 = 8.43 \text{ g/dm}^3$$
$$= 2.11 \text{ g per } 250 \text{ cm}^3$$

ACIDS, BASES AND SALTS—I 171

But as the given sample weighed 2.23 g

percentage of NaHCO$_3$ = $\frac{2.11}{2.23}$ × 100

= 94.6%

Note: This method assumes that all the impurities do not react with acid; if this is not assumed, then the titration gives the *total* alkali, and includes such things as sodium carbonate, etc.

Revision 18.2

1. Define the terms: end-point, molar solution, standard solution.
2. What are the functions of a burette, standard flask, and pipette?
3. Using the atomic weights on page 219 calculate the weights of the following required to neutralise 1 dm^3 of M KOH:
   Monobasic acids: butyric ($H_8C_4O_2$), lactic ($H_6C_3O_3$) and benzoic ($H_6C_7O_2$).
   Dibasic acids: tartaric ($H_6C_4O_6$) and sulphurous (H_2SO_3).
   Tribasic: phosphoric acid (H_3PO_4).
4. Calculate the molarities of sulphuric acid solutions in the following cases:

    25.3 cm^3 equivalent to 14.3 cm^3 1.0 molar alkali
    17.8 cm^3 equivalent to 23.6 cm^3 0.67 molar alkali
    28.3 cm^3 equivalent to 19.3 cm^3 0.74 molar alkali
    31.4 cm^3 equivalent to 25.6 cm^3 1.26 molar alkali

5. Calculate weights per 250 cm^3 to make the following solutions: 1.00M borax; 0.56M potassium carbonate; 0.90M oxalic acid; 5.04M sodium bicarbonate.

Chapter 19

Acids, Bases and Salts—II

In this chapter we shall consider some of the more important members of these classes, together with some combinations of them of particular bearing on bakery practice.

THE COMMON ACIDS

1. Sulphuric acid, H_2SO_4

This is the most famous and most important of all the acids. More sulphuric acid is made each year than any other acid, and it surpasses the other two 'mineral acids' (nitric and hydrochloric) partly because it is cheaper than either.

It is made by burning sulphur, or ores containing sulphur, in air. Sulphur dioxide is formed, and this is then oxidised further by more air in the presence of a catalyst, often vanadium pentoxide:

$$S + O_2 = SO_2$$
$$2SO_2 + O_2 = 2SO_3$$

This substance reacts instantly with water to form sulphuric acid:

$$SO_3 + H_2O = H_2SO_4$$

This is a colourless liquid, which is highly corrosive on the skin and will attack most metals. In addition to being an acid, it has a strong affinity for water in which it dissolves to form the dilute acid with evolution of much heat. It will often remove the elements of which water is made if they are present in compounds. Cane-sugar is converted into carbon, for instance, the sulphuric acid acting as a kind of catalyst:

$$C_{12}H_{22}O_{11} = 12C + 11H_2O$$

Sulphuric acid is used for making a vast number of other chemicals, particularly fertilisers, rayon, and the white pigment titanium dioxide. In the laboratory it is extensively used as a drying

ACIDS, BASES AND SALTS–II

agent, as a source of other chemicals, and for general laboratory purposes.

2. Hydrochloric acid, HCl

This is a solution of hydrogen chloride gas (HCl) in water (maximum concentration about 35% in 'concentrated hydrochloric acid'). It is made, commercially and in the laboratory, by combining hydrogen and chlorine gases obtained by electrolysis:

$$H_2 + Cl_2 = 2HCl$$

or by the action of concentrated sulphuric acid on common salt:

$$NaCl + H_2SO_4 = NaHSO_4 + HCl \quad \text{(in cold)}$$
$$\text{and} \quad 2NaCl + H_2SO_4 = Na_2SO_4 + 2HCl \quad \text{(heat)}$$

The gas is pungent and, because of its great tendency to dissolve in water, it fumes in a moist atmosphere, forming small droplets of the acid. Both gas and solution have a pungent smell, are colourless, and have a strongly acid taste. Hydrochloric acid is a typical strong acid and, unlike sulphuric and nitric acids, has few reactions not common to most acids. Very strong oxidising agents will, however, convert it into chlorine.

It is used for cleaning iron and in making chemicals in great variety. It also finds application in volumetric analysis and as a general purpose acid in the laboratory.

3. Nitric acid, HNO_3

Nitric acid is the remaining 'mineral acid' to consider. Most of it is now manufactured by a process in which ammonia is oxidised by air, catalytically, in several stages. Like the other two, it is a very strong acid and is corrosive to the skin and will attack most metals. It differs from most acids, however, in rarely giving hydrogen with metals. This is because, in addition to being an acid, it is also a powerful oxidising agent, and any hydrogen that might be formed is likely to be oxidised to water.

This is its most characteristic property. Indeed, one has only to heat it or expose it to light for some decomposition to occur, and the colourless liquid becomes yellow owing to the formation of nitrogen dioxide together with the oxygen:

$$4HNO_3 = 2H_2O + 4NO_2 + O_2$$

The same nitrogen dioxide is formed with copper and the concentrated acid at room temperature:

$$Cu + 4HNO_3 = Cu(NO_3)_2 + 2H_2O + 2NO_2$$

This gas has been used as a bleaching agent for flour.

In addition, nitric acid undergoes a certain type of reaction with organic compounds known as 'nitration', and this provides one of its most valuable functions, the resultant 'nitro-compounds' being immensely useful starting-points for making many other substances. Some are important in their own right, as the explosive T.N.T. Thus nitric acid has considerable military significance, and efforts to produce it economically have been particularly intense.

Nitric acid will attack most metals, but not gold. If, however, it is mixed with three times its own quantity of hydrochloric acid, then even gold will dissolve. The combination is known as 'aqua regia' ('royal water') for it can attack so 'noble' a metal. Its potency is believed to be due to a substance nitrosyl chloride (which has also been incorporated into some flour bleaches):

$$3HCl + HNO_3 = NOCl + Cl_2 + 2H_2O$$

Nitric acid is much employed in the laboratory, chiefly for making organic 'nitro-compounds' but also in the preparation of numerous simple nitrogen compounds as nitrogen dioxide.

4. Acetic acid

This is one of the simplest organic acids. Its formula, $H_4C_2O_2$, could, unfortunately, stand for numerous different compounds on account of the large number of ways in which these eight atoms can be arranged together. Therefore, organic chemists make it more definite by writing out a structural formula which shows not only what atoms are there, but also how they are joined together. Of the two structural formulae below, the second can be seen to be merely an abbreviation of the first:

$$\begin{array}{c} H \\ | \\ H-C-C=O \\ | \quad | \\ H \quad O-H \end{array} \quad \text{or} \quad CH_3.COOH$$

In future we shall represent it by the last formula. The dot stands for one valency bond.

Acetic acid is the chief component (apart from water) of vinegar. In the latter, it has resulted from the souring of wine in which the alcohol has been oxidised by the air under the catalytic influence of an organism known as *Mycoderma aceti*; in the equation below we also give the structural formula for alcohol (known to chemists as ethanol):

$$CH_3CH_2OH + O_2 = CH_3.COOH + H_2O$$

ACIDS, BASES AND SALTS—II

The acid is also made by a synthetic route from the gas acetylene, and little pure acetic acid is obtained from vinegar. Today it is chiefly produced by a variety of processes starting with petroleum.

Unlike the acids already considered, acetic acid is fairly weak, although typical in most of its reactions of acids as a whole. It gives hydrogen with many metals, is corrosive on the skin and liberates carbon dioxide from carbonates; it is strong enough to turn litmus a definite red. Although it is often met with as a colourless liquid, it will freeze on a cold day (18°C) to an ice-like solid, provided it is not diluted with water. For this reason, concentrated acetic acid is sometimes termed 'glacial' (or 'ice-like').

As a chemical intermediate, acetic acid is used in the preparation of aspirin and other medicinals. Directly or indirectly, it is employed in the preparation of a group of compounds known as acetates; amyl acetate, for example, has a taste and odour of pear-drops and is used as a synthetic flavouring, as well as a very widely met solvent; cellulose acetate is a synthetic fibre ('celanese') and the basis of non-inflammable films in photography.

Acetic acid is also generally used when a weak acid is required.

5. *Tartaric acid*, $H_6C_4O_6$

The structural formula for this may be written as:

$$O=C-C-C-C=O$$

with OH, H, H, OH substituents, or HOOC.CH(OH).CH(OH).COOH

This is a dibasic acid and forms two series of salts. With potassium hydroxide, for example, it gives potassium hydrogen tartrate, $KH_5C_4O_6$, and normal potassium tartrate, $K_2H_4C_4O_6$. The former of these occurs as 'cream of tartar', a discoloured deposit, during the fermentation of grape-juice to wine. From it the free acid may be prepared.

Tartaric acid is a white solid that is soluble in water. It is not poisonous and, therefore, is used as a constituent of a number of materials intended to be taken internally. Its role in baking powders will be considered later (p. 182); it is also mixed with bicarbonates in a number of effervescing drinks, when the two ingredients do not react until they are both dissolved. Tartaric acid is sometimes added in the preparation of certain confectionery.

176   PHYSICS AND CHEMISTRY OF BAKING

6. *Citric acid*, $H_8C_6O_7$
This tribasic acid has the structure:

$$\begin{array}{ccccccccc}
 & & \text{H} & & \text{OH} & & \text{H} & & \\
 & & | & & | & & | & & \\
\text{O}\!=\!\text{C}\!-\!&\text{C}&\!-\!&\text{C}&\!-\!&\text{C}&\!-\!&\text{C}\!=\!\text{O} \\
 & | & & | & & | & & | & \\
 & \text{HO} & & \text{H} & & \text{C} & & \text{H} & \text{OH} \\
 & & & & & /\!/\;\backslash & & & \\
 & & & & \text{O} & & \text{OH} & &
\end{array}$$

It occurs in lemon juice, but is usually made by a fermentation method from glucose. It is a white, soluble solid, and is often employed as an ingredient of icing-sugar and other foodstuffs.

Revision 19.1
1. Account for the importance of the three mineral acids in industry.
2. 'Acetic acid is a weak monobasic acid': explain what is meant by this statement.
3. How do acetic, tartaric and citric acids occur in nature?
4. Mention two laboratory uses, apart from volumetric analysis, of each of the three mineral acids.

THE COMMON ALKALIS

1. Sodium hydroxide, caustic soda, NaOH

The chief method of making this compound today is the electrolysis of brine (common salt solution). It is formed around the cathode and is not allowed to come into contact with the chlorine at the anode but is collected separately.

It is a white, very deliquescent solid, extremely soluble in water. It has a very caustic, or corrosive, action on most organic matter, including the skin. This is so because it is a very strong alkali. It has all the usual features of alkalis; being so strong, it will even react with the weakest acids, and will, for example, absorb carbon dioxide from the atmosphere to form sodium carbonate or bicarbonate:

$$2NaOH + CO_2 = Na_2CO_3 + H_2O \quad \text{(excess alkali)}$$
$$NaOH + CO_2 = NaHCO_3 \quad \text{(excess carbon dioxide)}$$

Caustic soda is the source of a considerable number of other sodium compounds. Included among these are detergents and soap. The latter is made from fats, of which the most important here is glyceryl stearate, present in mutton-fat, etc.; the reaction can be

ACIDS, BASES AND SALTS—II 177

expressed:

$$\text{fat} + \text{NaOH} = \text{soap} + \text{glycerol}$$

or,

$$\text{glyceryl stearate} + \text{NaOH} = \text{sodium stearate} + \text{glycerol}$$

More will be said about both fats and glycerol in a later chapter (p. 212).

2. Potassium hydroxide, KOH
Otherwise known as 'caustic potash', this substance is closely similar to caustic soda. It is, however, rather more expensive to produce, and therefore is met less often.

3. Calcium hydroxide, $Ca(OH)_2$
In contrast to potassium hydroxide, this alkali is the cheapest available on a large scale. It is formed by adding water to quicklime (calcium oxide) which, in turn, is made by heating limestone or chalk (calcium carbonate):

$$CaCO_3 \rightleftharpoons CaO + CO_2$$
$$CaO + H_2O \rightleftharpoons Ca(OH)_2$$

As the solid, it is usually met under the name 'slaked lime'. It is sparingly soluble in water, the solution being called 'lime-water'; if a lot of calcium hydroxide is added to water, a slurry or suspension results, known as 'milk of lime'. On heating, calcium hydroxide reverts to quicklime and water.

Lime-water is, of course, used in testing for carbon dioxide (p. 136). Like the caustic alkalis, it absorbs this gas from the air, though it is considerably weaker than they. Lime-water is sometimes added to flour to reduce its acidity. Slaked lime is a constituent of mortars used in building, and as a mild alkali generally in the laboratory.

4. Ammonium hydroxide, NH_4OH
Also a weak alkali, and weaker than calcium hydroxide, this is formed by passing the gas ammonia into water:

$$NH_3 + H_2O \rightleftharpoons NH_4OH$$

It is known only in solution, as, on warming the latter, it decomposes back into ammonia and water (hence the reversible arrow).

This solution, which always smells strongly of ammonia, is used for cleaning purposes, sometimes mixed with soap ('cloudy ammonia'). It also finds outlet in the preparation of ammonium salts and as a general-purpose weak alkali.

SOME COMMON SALTS

In this section only those salts which have a direct bearing on bakery processes will be considered.

1. Sodium chloride, NaCl

Otherwise known as 'common salt' or as 'salt' alone, this occurs in vast quantities in sea-water (around 3%). It is obtained from this source in hot climates when the heat of the sun can be used to drive off the water. In this country it is mined from deposits in the North West, where, in certain areas, it occurs in great quantities.

Much of the sodium chloride produced today is destined for production of other sodium compounds, particularly those which can be obtained from electrolysis of the aqueous solution. These include caustic soda and the products it forms with the chlorine that is evolved, notably sodium hypochlorite, a powerful bleach. Other sodium compounds (as soda) and sodium itself are made from salt, and the glazing of stoneware accounts for yet more.

The most familiar use is, of course, in foodstuffs. It is, in fact, essential to life, and even so-called 'saltless' diets involve the uptake of small amounts, chiefly in vegetable and meat products where it is always present to some degree. As an additive to flour it imparts both flavour and bloom to the finished loaf; it is one of the many substances capable of stabilising the gluten, and it also will retard the yeast action to some extent.

A test given by salt (and by all chlorides in solution) is the addition of silver nitrate and dilute nitric acid. A white precipitate is formed of silver chloride:

$$AgNO_3 + NaCl = AgCl \downarrow + NaNO_3$$

The nitric acid is there to prevent precipitation of many other silver salts at the same time; chlorides are the only salts to give a white precipitate under these conditions. (Bromides and iodides give precipitates, but of yellowish colour.)

2. Carbonates and bicarbonates

These are the salts of carbonic acid. As the latter is dibasic, it is possible to replace either both hydrogen atoms or one of them, giving the two series of salts:

$$H_2CO_3 \quad Na_2CO_3 \quad NaHCO_3$$
Carbonic acid — Sodium carbonate — Sodium bicarbonate

All give off carbon dioxide with acids. They differ in solubility and in the action of heat, however, and may be divided into four groups, shown in Table 19.1.

TABLE 19.1

Compounds	Solubility in water	Action of heat
Bicarbonates of sodium and potassium	soluble	Easily decomposed into carbon dioxide, water and the carbonate: e.g., $2NaHCO_3 = Na_2CO_3 + H_2O + CO_2$
Ammonium bicarbonate and carbonate	soluble	Easily decomposed into carbon dioxide, ammonia and water: e.g., $NH_4HCO_3 = NH_3 + CO_2 + H_2O$
Carbonates of sodium and potassium	soluble	Not decomposed (except for loss of water of crystallisation from $Na_2CO_3.10H_2O$)
Other carbonates	insoluble	Decomposed into oxide and carbon dioxide: e.g., $CaCO_3 = CaO + CO_2$

Only sodium, potassium and ammonium form well-defined bicarbonates out of solution, and 'ammonium carbonate' has a rather variable constitution.

Sodium bicarbonate is much used in baking powders (see p. 183), and because it is non-poisonous but mildly alkaline, is used as a medicinal antacid. When a very mild alkali is required, this compound is often the choice.

Although it is soluble in water, its solubility is not very great. Upon this fact depends the main method for its production. A solution of salt saturated with ammonia is treated with carbon dioxide; the ammonia, water, and carbon dioxide act as ammonium bicarbonate, and the latter reacts to form sodium bicarbonate and ammonium chloride. Of all these substances, the sodium bicarbonate is the least soluble and is precipitated from the concentrated

solution:
$$NH_3 + H_2O + CO_2 = NH_4HCO_3$$
$$NH_4HCO_3 + NaCl = NH_4Cl + NaHCO_3\downarrow$$

Potassium bicarbonate is much more soluble and cannot be prepared similarly. The alternative method is more expensive and the product less commonly met. It is, however, in all chemical respects similar to the sodium compound.

The carbonate and bicarbonate of ammonium are both made by interacting ammonia, carbon dioxide and water, and are both used in baking:

$$NH_3 + CO_2 + H_2O = NH_4HCO_3$$
$$2NH_3 + CO_2 + H_2O = (NH_4)_2CO_3$$

As mentioned above the carbonate is a solid of somewhat indefinite composition, and the formula $(NH_4)_2CO_3$ is an oversimplification. The crude salt is the essential constituent of *sal volatile* or smelling salts, a now unfashionable antidote to tendencies to fainting. As it decomposes, even in the cold to some extent, both ammonia and carbon dioxide are slowly evolved; the former stimulates the heart, the latter the breathing, hence perhaps the scientific basis of the mixture.

Sodium carbonate is manufactured in several ways, amongst them the action of heat on the bicarbonate:

$$2NaHCO_3 = Na_2CO_3 + H_2O + CO_2$$

It is a white crystalline solid, often met as $Na_2CO_3.10H_2O$, or washing soda; this loses its water of crystallisation on standing in the air by the process of efflorescence, and on heating goes rapidly to anhydrous salt; it does not decompose any further.

It is much more strongly alkaline than the bicarbonate, and is a primary standard for volumetric analysis. It is sometimes used as a substitute for caustic soda (e.g., in soap-making) and is employed in the manufacture of glass, where a mixture of sand, limestone and soda is fused together. Because of its alkaline action it has detergent properties, hence the use of the decahydrate as 'washing soda'.

Potassium carbonate, K_2CO_3, is similar to sodium carbonate.

Of the other carbonates, the only one that need detain us is calcium carbonate, $CaCO_3$. This occurs naturally in vast deposits as chalk, limestone, and other minerals. Its importance in the formation of hardness in water has already been discussed (p. 155).

When limestone or chalk are heated fairly strongly, carbon dioxide is lost and quicklime (calcium oxide) is formed:

$$CaCO_3 = CaO + CO_2$$

This is the basis of 'lime-burning' which has been carried on for centuries in 'lime-kilns'. The quicklime may then be 'slaked' to form calcium hydroxide. Much limestone is, of course, used as a building material, while this is also true of marble, another form of calcium carbonate.

Marble is often one of the reagents in the laboratory preparation of carbon dioxide (p. 135). Calcium carbonate is often employed as a 'filler' to various substances, including flour.

3. Sulphates

These are the salts of sulphuric acid. Bisulphates are known (*cf.* bicarbonates) but do not concern us here except perhaps to note that most lavatory cleaners owe their potency to sodium bisulphate, $NaHSO_4$.

All sulphates that are soluble give a precipitate of barium sulphate when their solutions are mixed with barium chloride solution and hydrochloric acid, e.g.,

$$CuSO_4 + BaCl_2 = BaSO_4 \downarrow + CuCl_2$$

Calcium and magnesium sulphates occur naturally and cause permanent hardness in water (p. 156); naturally occurring forms include $CaSO_4.2H_2O$, gypsum, and $MgSO_4.7H_2O$, Epsom salts. Gypsum, on heating, forms plaster of Paris, and the medicinal use of Epsom salts is well known. Magnesium sulphate is an improver sometimes added to flour to retard the action of the diastase (p. 210) and prevent the gluten from breaking down during the fermentation. Calcium sulphate is sparingly soluble, magnesium sulphate quite soluble in water.

Sodium and potassium sulphates are also soluble crystalline solids. The former is, like magnesium sulphate, a mild purgative, and can be an additive to flour. It also finds application in the manufacture of glass and paper.

4. Nitrates

Nitrates, salts of nitric acid, are all soluble in water. Sodium nitrate is sometimes employed as a preservative for tinned meat, but is most used as a fertiliser and in the production of potassium nitrate which, in turn, is a constituent of gunpowder and fireworks. This also has been used for preserving meat.

5. Phosphates

Phosphoric acid exists in a number of forms, of which the commonest is H_3PO_4, a tribasic acid. Together with its salts, it is of the greatest importance for development of life, and in particular as a fertiliser.

The phosphates of calcium are probably the most important. The salt, with all hydrogen replaced by calcium, is $Ca_3(PO_4)_2$; this occurs as a mineral, phosphorite, and also in bones. Unfortunately, the naturally occurring calcium phosphate is practically insoluble in water, and for this reason cannot be assimilated by plants from the soil. If, however, it is treated with acid to form an acid phosphate $Ca(H_2PO_4)_2$, it now gives a soluble fertiliser, of great value. The cheapest, and therefore commonest, treatment is to treat it with concentrated sulphuric acid, the resulting mixture of acid calcium phosphate and calcium sulphate being known as 'superphosphate':

$$Ca_3(PO_4)_2 + 2H_2SO_4 = Ca(H_2PO_4)_2 + 2CaSO_4$$

This same acid calcium phosphate is also used in bakery. Because of its acid properties (it still has replaceable hydrogen atoms) it will liberate carbon dioxide from bicarbonates and is, therefore, used in some baking powders (especially in the USA). It also finds application as an 'improver' to flour, for it helps to stabilise the gluten and at the same time nourishes the yeast, which cannot live without phosphates. Yet again, it is mixed with benzoyl peroxide in one of the methods for bleaching flour. Other phosphates also are used as flour improvers.

In all these applications it has the advantages of being cheap, not so strongly acidic as to be harmful, and it is also nutritious both to the micro-organisms causing fermentation and to the human body. It is prepared for these purposes by the action of phosphoric acid on normal calcium phosphate or on limestone.

Other phosphates met in the laboratory are less important for bakery practice. They include the three sodium salts derived from 'ortho'-phosphoric acid:

$$Na_3PO_4, \quad Na_2HPO_4, \quad NaH_2PO_4,$$

the middle one being the commonest. We have already mentioned sodium hexametaphosphate, used in water-softening (p. 159); this can be made from the last of the sodium compounds mentioned above by strong heating followed by rapid cooling:

$$6NaH_2PO_4 = (NaPO_3)_6 + 6H_2O$$

6. Tartrates

These are the salts of tartaric acid (p. 175). The chief members are these:

Potassium hydrogen tartrate, 'cream of tartar', is formed as a by-product in the fermentation of grapes, when it occurs as a

ACIDS, BASES AND SALTS—II 183

sediment in the casks. It has an acid taste, and is used in dyeing, in medicine and in baking powders (p. 184).

Sodium potassium tartrate, is also known as 'Rochelle salt', and is made by treating cream of tartar with sodium carbonate or hydroxide:

$$KH_5C_2O_4 + NaOH = NaKH_4C_2O_4 + H_2O$$

It is used in Fehling's solution in organic chemistry (a useful oxidising agent also containing alkali and copper sulphate); medicinally, it is a laxative and is incorporated in several well-known preparations of the effervescent kind.

COMPOSITION OF BAKING POWDERS

All these are based upon bicarbonates or carbonates as the source of the carbon dioxide needed to swell the gluten. They are, of course, alternative to yeast; baking powders have several strong advantages over fermentation methods, not least of which are the following:
(i) Speed: Baking powders do not require a preliminary fermentation period before baking, and thus save time.
(ii) Conservation of carbohydrate: Carbohydrates are converted into alcohol and carbon dioxide during fermentation, and thus are lost to the final product. With baking powder no inroad at all is made into carbohydrates present.
(iii) Predictability: Yeast is always, even at best, a slightly variable substance, and one cannot with any degree of confidence predict exactly how much carbon dioxide a given amount will produce. But baking powders give reproducible results, and one can calculate with precision the quantity of gas obtainable.
In general, baking powders gain on grounds of both economy and convenience.

We now consider some common baking powders:

1. Sodium bicarbonate alone

Sodium bicarbonate (bicarbonate of soda) is obtainable in high purity (*cf.* p. 179), and gives carbon dioxide merely on heating:

$$2NaHCO_3 = Na_2CO_3 + CO_2 + H_2O$$

Unfortunately, this most simple of all methods suffers from the production also of the carbonate which is quite strongly alkaline. Its presence is undesirable because of the unpleasant taste, and because the alkali may destroy certain vitamins and lead to some discolouration.

2. Sodium bicarbonate and tartaric acids

In view of what has just been said, it is apparent that carbon dioxide must be produced from the bicarbonate by other means than just heating. The obvious method is to add an acid. Clearly, the acid we choose must be neither poisonous nor corrosive (that excludes the mineral acids), and it must be free from strong and unpleasant flavour. Moreover, it should not react the instant the dough is mixed, but must be able to be restrained in its action until the mixture is in the oven and heat applied. This last requirement necessitates a solid acid.

Tartaric acid by itself has been suggested:

$$2NaHCO_3 + H_6C_4O_6 = Na_2H_4C_4O_6 + 2H_2O + 2CO_2$$

The di-sodium salt is an aperient, however, and its occurrence in appreciable amounts in the product would be rather disastrous. Cream of tartar, however, is an acid which does not produce this substance, but the milder Rochelle salt:

$$KH_5C_4O_6 + NaHCO_3 = NaKH_4C_4O_6 + H_2O + CO_2$$

A little tartaric acid may be added as the cream of tartar is not very soluble in cold water and so may not start to act early enough in the process. But the combination cream of tartar + sodium bicarbonate is still one of the most important.

The equation above reveals an important fact about the reaction, namely, that the molecules of the reagents are in a ratio of 1.1; hence a mole of one requires a mole of the other. As the molecular weights of the bicarbonate and tartrate are respectively 84 and 188, their weights in the reaction must be in the ratio of 84:188, or 1:2.24. Hence they must always be mixed in the proportion of about one to two parts by weight.

3. Sodium bicarbonate and acid calcium phosphate

Acid calcium phosphate (p. 182) has been used as an alternative to the tartaric acids, especially in the USA. It reacts as follows:

$$Ca(H_2PO_4)_2 + 4NaHCO_3 = Ca(Na_2PO_4)_2 + 4H_2O + 4CO_2$$

4. Ammonium carbonate or bicarbonate

These salts (p. 180) may be used alone, as they break down on heating into gases only, leaving no objectionable solid residue of any kind:

$$(NH_4)_2CO_3 = 2NH_3 + CO_2 + H_2O$$
$$NH_4HCO_3 = NH_3 + CO_2 + H_2O$$

They are used in biscuit manufacture.

ACIDS, BASES AND SALTS—II 185

We have already spoken of the reliability of the baking-powder methods for producing known quantities of carbon dioxide. How may these quantities be known? The estimation of carbon dioxide may be of two types. We may estimate it by volume or by weight. Suppose exactly one gram of the solid is started with, and it is desired to find out how much CO_2 it will produce. In the first case the volume of gas would be measured, and in the second its weight.

VOLUMETRIC ESTIMATION OF CARBON DIOXIDE

The apparatus used is illustrated in Fig. 19.1. The flask contains an accurately known weight of sample. The funnel contains more than sufficient dilute acid (e.g., sulphuric) to liberate all its CO_2. The solution in the glass manometer is brine saturated previously with carbon dioxide; if it were not saturated, some of the gas produced would dissolve and would not be indicated in the volume change.

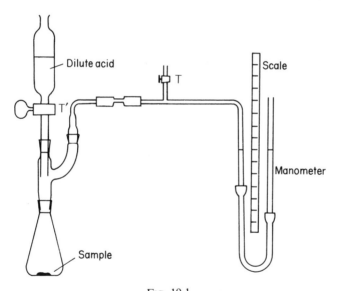

FIG. 19.1

The levels in the two limbs of the manometer are equalised by opening the tap T, which is then closed again. The acid is now run in as quickly as possible, and the tap T' immediately closed.

As the gas is evolved, the level in the left-hand tube descends, and that in the right-hand tube rises. When there is no further change, the right-hand tube is lowered until the levels are once more the same in the two tubes, but lower than at first. This means that the pressure inside is the same as that of the atmosphere. The difference between the old and new levels is thus the volume of gas produced. (For very precise values one would have to deduct the volume of liquid added, but this will be relatively small.) It is a simple matter to calculate how much gas would be evolved from 1.00 g of powder.

GRAVIMETRIC ESTIMATION OF CARBON DIOXIDE

The gas evolved from a known weight of sample is passed through weighed bulbs of 50% potash solution. These are re-weighed after reaction, and the increase is the weight of gas absorbed. Precautions are taken to prevent CO_2 from the air from entering at either end of the line (hence the soda-lime tubes), and the gas is dried with concentrated sulphuric acid to prevent any water vapour from entering as this might also increase the weight of the potash tubes. The condenser, with cold water circulating, is to return as much water as possible to the flask by condensing out the steam.

A better method is to produce the carbon dioxide by the action of an acid on the sample. In this case, the apparatus to the left of the sulphuric acid bottle in Fig. 19.2 is replaced by a generator flask, as in Fig. 19.1. It may also be more convenient to replace the potash bulbs with U-tubes containing soda-lime. The volume of gas evolved should be twice that obtained by merely heating the same weight of sample.

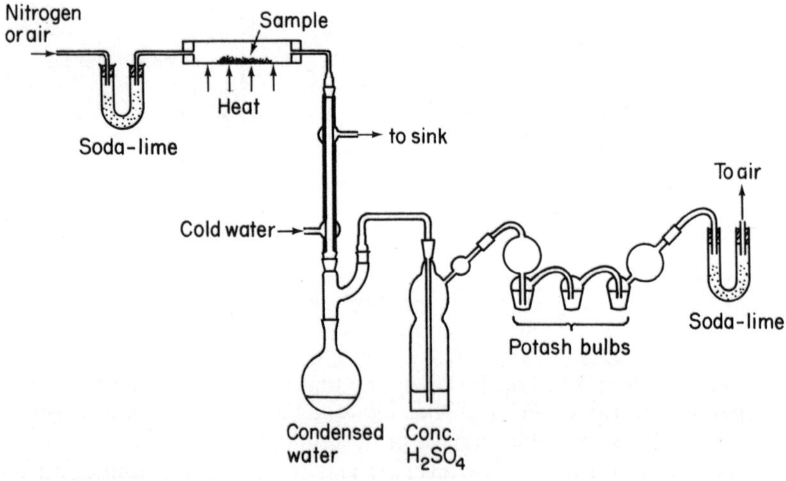

FIG. 19.2

ACIDS, BASES AND SALTS–II 187

The apparatus in Figs. 19.1 and 19.2 has been shown with old-fashioned large-scale components for the sake of clarity. In practice today, however, one would almost always use semi-micro or even micro techniques which use far less materials and are much faster.

Revision 19.2

1. Write equations for the action of heat on limestone, slaked lime, potassium bicarbonate, washing-soda, ammonium carbonate.
2. What are the industrial uses of caustic soda, calcium hydroxide, calcium phosphate, cream of tartar?
3. Write a short account of the composition of baking-powders.
4. How could you distinguish between sodium bicarbonate and carbonate; sodium hydroxide and carbonate; calcium chloride and magnesium sulphate; calcium and potassium carbonates?

Chapter 20

Oxidation

NATURE OF OXIDATION

Oxidation is a process that is much more far-reaching than seems at first sight. The following definition is only a partial one, but it is true as far as it goes and will suffice to begin with:

Oxidation—the addition of oxygen to a substance or the removal of hydrogen from it.

To this we must add the definition of reduction:

Reduction—the addition of hydrogen to a substance or the removal of oxygen from it.

Other related definitions are:

Oxidising agents—substances that cause oxidation of others.

Reducing agents—substances that cause reduction of others.

Whenever oxidation takes place, reduction does as well. This is obvious if we regard these processes as transfers of either oxygen or hydrogen. Consider the reaction between coke and steam:

$$C + H_2O = CO + H_2$$

Oxygen has passed from the water to the carbon. Therefore, the carbon has been oxidised (to the monoxide), and steam has been reduced (to hydrogen). The steam is an oxidising agent in this example, and the carbon a reducing agent. The overall reaction could be equally well called oxidation of carbon or reduction of steam.

It is, however, possible to be even more general than this using the concept of *oxidation* number. In very broad terms this may be said to equal the valency of the element in any given compound, with the additional proviso that, in binary compounds, metals (and hydrogen) usually have positive numbers, and non-metals with which they are combined have negative numbers. The *total* must always be zero. Some examples are given below:

Compound	Oxidation numbers
NaCl	Na ($+1$) ; Cl (-1)
$CaCl_2$	Ca ($+2$) ; Cl (-1)
Al_2O_3	Al ($+3$) ; O (-2)
ZnS	Zn ($+2$) ; S (-2)

Where compounds between two non-metals are concerned the same rules apply, but which element receives the negative number is determined by general chemical considerations; thus in the combination between non-metals and oxygen, the oxygen is always negative. Free elements are given oxidation numbers of zero. The following are further examples:

Compound	Oxidation numbers
H_2O	H ($+1$) ; O (-2)
CO_2	C ($+4$) ; O (-2)
NCl_3	N ($+3$) ; Cl (-1)
CCl_4	C ($+4$) ; Cl (-1)
Cl_2O	Cl ($+1$) ; O (-2)

This brings us to a new definition of oxidation. An element is said to be oxidised if its oxidation number has been increased. Thus in the reaction between steam and coke:

$$C + H_2O = CO + H_2$$

oxidation numbers are:

	before reaction	after reaction
carbon	(0)	($+2$)
oxygen	(-2)	(-2)
hydrogen	($+1$) × 2	(0)

Hence carbon has been oxidised. Now consider what happens when methane is treated with chlorine. Many different reactions can take place including the following:

$$CH_4 + 4Cl_2 = CCl_4 + 4HCl$$

	before reaction	after reaction
hydrogen	($+1$) × 4	($+1$) × 4
chlorine	(0)	(-1) × 8
carbon	(-4)	($+4$)

so in this case chlorine has oxidised the methane though no oxygen is involved.

Revision 20.1

1. Define oxidation, reduction, oxidising agent, oxidising number.
2. Assign oxidation numbers to each of the elements in the following: Cl_2, CaF_2, H_2S.
3. Which element (if any) is oxidised in the following changes?

$$2H_2 + O_2 = 2H_2O$$
$$2SO_2 + O_2 = 2SO_3$$
$$NH_3 + 3Cl_2 = NCl_3 + 3HCl$$
$$NaOH + HCl = NaCl + H_2O$$
$$CuO + H_2 = Cu + H_2O$$
$$CH_4 + 2Cl_2 = C + 4HCl$$
$$CuSO_4 + Zn = Cu + ZnSO_4$$

We shall now proceed to consider some important processes in which oxidation occurs.

CORROSION OF METALS

Everyone is familiar with the way some kinds of iron readily rust away. Equally well known is the resistance of the so-called 'noble' metals, such as gold and silver, to any kind of chemical attack. Unfortunately, iron is much more useful for most constructional work than gold and silver, and so the problem is how to protect iron and similar useful metals from this destruction by corrosion.

Corrosion is, essentially, an oxidation by atmospheric oxygen. It seems to require traces of moisture and perhaps carbon dioxide, and arises from the formation on the metal surface of tiny electrical cells. Certainly, impurities of a kind that could form part of such cells greatly accelerate the rates of corrosion. What remedies can be found for this state of affairs? The following are the chief ways in which corrosion can be avoided.

1. Covering the metal with non-metallic protection: Air cannot attack a metal if it is not in contact with it; therefore, the obvious thing to do is to cover the surface with a layer of something the air cannot penetrate. This is the basis of this and the next two methods.

Painting the metal is the simplest example, though not usually applicable when food is involved. Enamels may be baked on, and these provide excellent protection, but they tend to be brittle and, once a crack appears, corrosion proceeds apace. For baking-tins, it is common to apply a silicone fluid, allow to dry and then heat; this forms a glossy surface which protects against corrosion, and also dispenses with the need for grease in baking. Similar ability to obviate greasing exists in the PTFE (polytetrafluoroethylene) films used on metal equipment in some bakeries.

2. *Plating with 'noble' metals:* Most of the possible 'noble' metals that withstand corrosion are too expensive or otherwise unsuitable for use by themselves. But if they are plated on iron or steel, much smaller costs are involved and the advantages of the underlying metal are retained. Silver and gold plating are familiar in many articles. Cutlery bearing the letters EPNS is 'electroplated nickel silver'. Chromium plating is also a familiar feature of life today; the resistant metal chromium is generally plated on to nickel, which has been plated on to steel; the nickel enables both chromium and silver to adhere more firmly.

From the viewpoint of the food industries, however, there is no doubt that the tin-plated iron combination is the most important. Tin is melted and iron dipped into it, the surplus tin removed by rollers, and the tin-coated metal allowed to cool. Electrolytic methods for depositing the tin have been used, but all types of tin-plate depend upon the resistance of tin to corrosion by the atmosphere for their continued brightness and lustre.

3. *Plating with reactive metals:* A method for protection of iron in particular is to coat the surface with zinc; this can be effected by dipping the cleaned metal in molten zinc (galvanising), or by spraying the zinc on to it (sherardising). In either case, the zinc layer protects the iron because it is itself preferentially oxidised by moist air; in the presence of zinc, in fact, iron is much harder to oxidise than in its absence. This is sometimes known as 'sacrificial corrosion', the zinc being 'sacrificed' to save the iron. Galvanised bins of various sorts are widely used, and it is important to know that, if the surface is broken at any point, the exposure of zinc and iron together in the presence of moisture accelerates the rate at which the zinc corrodes, and, therefore, the rate at which the iron surface is exposed. Careful handling is essential for galvanised ware.

4. *Addition of alloying elements:* Suppose it is required to ensure freedom from corrosion for an iron object. Instead of coating it in some way, we may add to the molten metal (before it is fashioned into the final shape) small amounts of other metals which will produce an alloy of iron which will resist corrosion much better than the pure iron. This is particularly important when the iron has already been converted to steel, which has a small percentage of carbon (usually 0.15 to 1.5%).

In the bakery we constantly use stainless steel equipment. This consists of ordinary steel to which has been added up to 20% of chromium and a little nickel. Other steel alloys are widely used on account of their hardness and resistance to abrasion, as well as their incorrodibility. These include those formed by addition of such elements as manganese, vanadium, and tungsten.

5. *Employment of resistant metals:* Instead of coating or alloying a metal as iron, we may dispense with this metal and use an entirely different one where problems of corrosion are largely non-existent. This, of course, is an ideal solution, but it has sometimes been found to be a practicable one as well. The outstanding example is the use of aluminium, a metal which has many advantages over iron in any case, apart from its resistance to corrosion.

Aluminium is much less dense ('lighter') than iron, and therefore vessels and other objects of the same volume weigh considerably less in aluminium than in most other metals, iron included. The densities of some of these relative to aluminium (relative density = 1.0) are: steel, 2.9; brass, 3.1; copper, 3.3; lead, 4.2. This lightness is obviously of great advantage in the aircraft industry and aluminium and its alloys have been extensively used there. It also finds applications in many other situations where lightness is convenient, if not essential. To reduce to a third of their weight objects such as saucepans and baking tins that are constantly being lifted about is clearly a good thing.

Cooking vessels are often made of aluminium for yet a third reason, and that is that the metal has a very high thermal conductivity; hence the contents are heated more rapidly than with other metals, with economy in time and fuel. With vessels of this material it is important to remember that alkalis such as washing soda must never be used as cleaning agents because aluminium is attacked to form aluminates and hydrogen: e.g.,

$$2Al + Na_2CO_3 + 3H_2O = 2NaAlO_2 + 3H_2 + CO_2$$

The reason for aluminium's resistance to corrosion is not just unreactivity. It is, in fact, a most reactive metal. The crux of the matter is that a freshly exposed surface of aluminium becomes immediately oxidised by the air to a film of oxide, and this film then adheres most firmly to the underlying metal, so protecting it from further corrosion:

$$4Al + 3O_2 = 2Al_2O_3$$

We are thus back to the first methods of protection—the provision of protective coatings, only in this case they appear automatically. It is, indeed, possible to thicken the film by electrolytic means when the aluminium is made the anode of an electrolytic cell and the oxygen evolved there attacks the metal to a greater extent than

OXIDATION 193

does the air. 'Anodised' aluminium has the additional advantage that the film can be readily stained and thus made to take on many attractive colours. Normally the coating is so thin that it is quite invisible. Once it is removed corrosion is rapid, and this can happen on contact with mercury or immersion in sea-water.

Other resistant metals are of considerably smaller importance for our purposes. The 'coinage metals' (so called from their most commonplace use in years past), copper, silver and gold have very high chemical inertness, especially the last two. Silver also has the highest thermal and electrical conductivity of all metals; copper is next.

Revision 20.2
1. Give three methods for protecting iron from corrosion.
2. Give two advantages and one disadvantage in the use of aluminium for baking-tins.
3. Explain the following terms: galvanising, sacrificial corrosion, tinplate, anodising, EPNS

COMBUSTION OF FUELS

The burning of a fuel is always an oxidation reaction in which the fuel is oxidised by air (or sometimes pure oxygen), and heat is evolved. Not all fuels, of course, involve combustion; atomic power does not involve it at any stage, and electricity heats without any chemical reaction taking place, although it may have been generated by steam power which, in turn, depends upon burning of some secondary fuels.

We shall consider here a few of the most common fuels from the point of view of their chemistry. We are not greatly concerned to show their rival merits, as these often depend upon quite nonchemical points of economics and costing.

1. GASEOUS FUELS

All common gaseous fuels contain hydrogen, carbon monoxide, or hydrocarbons; the last are compounds of hydrogen and carbon only, and include hundreds of members all of which can be represented by the general formula C_xH_y, where x and y are integers. The oxidations of these can be expressed by the equations:

$$2H_2 + O_2 = 2H_2O$$
$$2CO + O_2 = 2CO_2$$
$$2C_xH_y + \tfrac{1}{2}(4x+y)O_2 = 2xCO_2 + yH_2O$$

(The last equation assumes that there is plenty of oxygen; if not, carbon monoxide will be formed instead of the dioxide.)

(a) Natural gas

For a century and a half town gas has been made by roasting coal in the absence of air to about 1000 C; the chief constituents were 52% hydrogen, 32% methane (CH_4) and up to 9% of the poisonous carbon monoxide (CO). Since the 1970s, however, it has been rapidly replaced by natural gas from under the North Sea or other deposits of petroleum. This gas consists largely of methane and has about twice the heating power of coal gas and none of its poisonous qualities. It burns to form carbon dioxide and water only.

Two practical points emerge from the chemistry of natural gas. One is that a plentiful supply of air must always be present. Otherwise there may not be enough to oxidise completely all the methane and the dangerously poisonous carbon monoxide will be produced. This is, unfortunately, almost odourless, and it may go undetected until it causes severe poisoning in persons near the apparatus. Good ventilation is obviously the answer.

This also resolves the other main difficulty in using natural gas. A glance at the equations below will show that methane burns to form water vapour. In an ill-ventilated bakehouse or kitchen, condensation on cool surfaces is a real problem. But if extractor fans expel the water vapour almost as soon as it is formed this difficulty also disappears.

$$CH_4 + 2O_2 = CO_2 + 2H_2O$$
$$2CH_4 + 3O_2 = 2CO + 4H_2O$$

We shall now look at some other gaseous fuels, rather less important from our point of view.

(b) Compressed hydrocarbons

The hydrocarbons propane, C_3H_8, and butane, C_4H_{10}, are obtained to-day in large amounts from petroleum refining. Although they are gases at room temperature and pressure, on compression they readily liquefy and may be stored in cylinders, etc. Such fuels are widely used where no mains gas supply is available, under such names as 'calor gas', 'propagas', etc. A valve in the container on being opened releases the pressure and some liquid turns into the inflammable gases which are then burnt in the same kind of way as town-gas. Recently, liquefied methane (CH_4) has become important although only on the industrial scale.

(c) Water-gas

This is important industrially both as a gaseous fuel and as a reducing agent. It is made from water (hence its name) by passing steam over red-hot coke; both components are capable of oxidation:

$$H_2O + C = H_2 + CO$$

As this reaction absorbs heat, the temperature of the coke gradually falls. Therefore, it is from time to time raised again by shutting off the steam and substituting air, when the coke burns in the usual way and heats up again.

(d) Producer gas

This gas is made by passing limited amounts of air over heated coke; the oxygen is converted into carbon monoxide (with some dioxide), and the nitrogen is unchanged:

$$2C + O_2 = 2CO$$

Containing so much nitrogen, at least 80%, producer gas is a poorer fuel than the others mentioned above.

(e) Acetylene

Although a hydrocarbon (C_2H_2) this is not compressed like propane, etc., because it is liable to explode if this is done. It is dissolved in acetone, however, and is of great value in welding, when it is mixed with air or oxygen in the oxy-acetylene torch. It is made from petroleum.

2. LIQUID FUELS

Consider the following list: petrol (gasoline), paraffin (kerosine), diesel fuel, jet fuel, lamp oil. All these liquid fuels are essentially hydrocarbons; more than this, they are all of the same hydrocarbon family known as the alkanes. Not only can they be written as C_xH_y, but we can add that $y = 2x + 2$, so their formula becomes C_xH_{2x+2}. In fuels many other substances are added to some of them, but they remain basically mixtures of this type of hydrocarbon. All are prepared from nature's storehouse of such compounds, petroleum (remains of long-perished marine creatures). It is the abundance of these deposits that makes petroleum products as the above so important. Only as the 1980s come upon us is there a general realisation that such deposits will not last for ever, so responsible use of these resources is essential.

3. SOLID FUELS

The traditional fuel in Great Britain is coal. Very complex in nature, it is the remains of fossilised vegetation of former ages. Contrary to popular belief, it contains little free carbon, but is rich in the element in a combined form. It has an important disadvantage in the character of its oxidation products. On heating, it readily forms volatile materials, not all of which may be burnt. Hence a smoky flame is produced, and corrosive tar is liable to accumulate in chimneys. Also, sulphur-containing compounds in the coal burn to sulphur dioxide which is toxic and which, eventually, is further oxidised to sulphuric acid, so adding to the corrosive effects of the polluted air. In 'smokeless zones' coal cannot be used, at least in the traditional manner.

When coal is carbonised (heated in the absence of air), there remains an involatile residue known as coke, which is a rather impure form of carbon. When it burns in air, it does not give off volatile tars or smoke, and is thus cleaner and less harmful than coal. Similar to coke in this last respect is the naturally occurring coal known as anthracite.

Although coke does not produce volatile matter as does coal, it often has a bluish flame over its surface as though some such material had been formed and was burning in the air. The explanation for this is that in the body of the hot fuel there is a relatively poor supply of air; hence carbon monoxide, not dioxide, is produced; but when the latter reaches the surface of the coke, there is more air and the monoxide now burns. Thus it is not a volatile substance from decomposition of the fuel, but a gas formed by its partial oxidation:

$$2C + O_2 = 2CO$$
$$2CO + O_2 = 2CO_2$$

Of the other solid fuels, only charcoal needs to be mentioned. This again is rather impure carbon and stands in the same relation to wood as coke does to coal, being made by heating wood in very restricted supplies of air. It is used as a fuel in barbecues, but no longer is very significant as a general means of heating. Being a good adsorbent, however, it is often employed in purifying liquids and gases from which it adsorbs impurities, e.g., it decolourises the crude sugar solutions that would otherwise crystallise to brown sugar.

Both charcoal and coke have been much used as reducing agents in the extraction of metals from their oxide ores: e.g.,

$$C + ZnO = Zn + CO$$

Coke has almost entirely superseded charcoal for these purposes.

Revision 20.3

1. Name the inflammable components of 'calor gas', water gas, producer gas, and give equations for their complete oxidation.

OXIDATION 197

2. What products are formed from the carbonisation of coal?
3. Explain the need for adequate ventilation in bakehouses having gas-fired ovens.
4. Briefly account for the advantages of coke over coal as a fuel.

BLEACHING OF COLOURED SUBSTANCES

Many coloured organic compounds may be easily converted to colourless materials by the process of oxidation. Because of this, most bleaches are strong oxidising agents, capable of removing many stains and unwanted colours when these are due to organic molecules. In some cases it is possible to reverse the bleaching action and to restore the colour by using strong reducing agents.

In this section we give some of the commonest bleaches, bearing particularly in mind the problem of bleaching flour.

1. CHLORINE
The element chlorine will bleach flour, coloured fabric, hair, and many other materials if traces of moisture are present. The bleaching action seems to be due to hypochlorous acid (HClO) formed thus:

$$Cl_2 + H_2O = HCl + HClO$$

The hypochlorous acid bleaches by releasing its oxygen and becoming itself reduced to HCl.

Pure chlorine is extremely dangerous, and has been used as a military gas; also it is highly reactive to most organic substances, and its use as a bleach is out of the question in practically all circumstances. It is a gas, however, and so can be diluted many hundreds of times with air, and under these conditions it has been used to bleach flour. It is sometimes mixed with 0.5% of nitrosyl chloride (NOCl), and the mixture is known as beta gas.

2. HYPOCHLORITES
When chlorine reacts with alkalis, one of the products will be a hypochlorite, e.g., sodium hypochlorite (NaClO):

$$Cl_2 + 2NaOH = NaCl + NaClO + H_2O$$

This substance, known only in solution, is a powerful oxidising agent and is the basis of many domestic bleaches. It also acts strongly as a germicide. Being available only in solution, even a dilute one, it is not applicable as a bleach for flour. It is, however, valuable for general disinfection and as bleach for cloths.

3. Ozone

This is a form of oxygen, and has three atoms to the molecule, O_3. It is formed when electricity is allowed to leak through ordinary oxygen without formation of a spark (which would destroy the ozone). This 'silent electric discharge', as it is called, produces a gas with a few per cent of ozone. It is used today chiefly as a purifier of the air, as it also destroys bacteria. It was, however, once used as a bleach for flour, but is not very satisfactory.

4. Nitrogen Dioxide

This gas, miscalled nitrogen peroxide, is met as NO_2 or as N_2O_4, and is produced by the action of copper or other metals on nitric acid (p. 173), and in other ways. It is a very drastic bleach, but has been used for flour in small amounts.

5. Nitrogen Trichloride

More flour was formerly bleached by this substance in the UK than by any other. It is a volatile liquid made by the action of chlorine on ammonia or ammonium chloride solution:

$$NH_4Cl + 3Cl_2 = NCl_3 + 4HCl$$

The pure substance is highly explosive, but is never required for bleaching. For this purpose, a little of the vapour is carried in a stream of moist air.

This substance, known as 'agene', some years ago suffered a setback when it was discovered that flour bleached with it could cause hysteria in dogs. Although it appeared to be harmless to man, its use has now been discontinued.

6. Chlorine Dioxide

This gas (ClO_2), made by the action of acids on chlorates or by passing chlorine into sodium chlorite solution, has been widely used as a bleach for flour.

7. Benzoyl Peroxide

This is an organic oxidising agent, $(C_6H_5CO)_2O_2$, and also a solid. It has been used as a bleach for flour in admixture with a number of other solids as calcium phosphate, alum, calcium carbonate, starch, etc.

Two other oxidising agents, though not bleaches themselves, have been added as 'improvers' to flour: ammonium persulphate $(NH_4)_2S_2O_8$, and potassium bromate, $KBrO_3$. Their oxidising action appears to link together strands of molecules in the gluten, so giving rise to a 'stronger' dough. Both are decomposed by heat.

Revision 20.4
1. What are 'agene' and 'beta gas'?
2. Give the action of chlorine on water, caustic soda, ammonium chloride.

Chapter 21

Basis of Organic Chemistry

What is life? This question has haunted men in the past and is likely to go on doing so despite much of what science has discovered in recent years. Whether or not we shall ever have the complete explanation of biological life, one chemical fact is quite clear: it is impossible to have living matter as we know it without the element carbon. The chemical composition of animals and plants is immensely complicated but every living cell and organism has this simple feature in common, and all are built up of carbon compounds. (It is, of course, untrue to say that everything containing carbon is, or was, living.)

We may well ask why carbon should be so essential to life, and the short answer to this is just that carbon alone of all the 100 or so elements can form extremely complex compounds, and complexity does seem necessary for life to exist. In particular, proteins, which are present in all living cells, are highly complicated molecules. How it is that carbon is able to form such compounds we shall shortly consider.

The chemistry of living organisms was once termed 'organic chemistry', but as a matter of convenience, this term is now defined as the chemistry of carbon compounds (excluding a few very simple members as the oxides of carbon, carbonates and one or two others). This redefinition is desirable because it includes great numbers of compounds which have not so far been found to have any connection with life but which, chemically, are closely related to those that do. It is thus a wider definition than the older one.

Chemical compounds are thus divided into two groups: organic, as thus defined, and inorganic. The latter includes all compounds not included in the former. We thus have the chemistry of carbon compounds on one hand and that of compounds derived from all other elements on the other.

This arrangement seems hopelessly one-sided until one realises that the total number of organic (carbon) compounds known to date far surpasses that of all the other elements put together. And

BASIS OF ORGANIC CHEMISTRY

this brings us to the heart of the matter when we enquire why it is that carbon behaves in this unique way.

Carbon is the only element capable of combining with itself to form chains of any number of atoms

All other elements are incapable of behaving like this. They may form quite stable bonds with elements different from themselves, but none gives a stable arrangement when joined on to another atom of itself. Thus the unit that we are considering could be written thus:

$$\begin{array}{c|c} | & | \\ -C\!\!-\!\!\!-\!\!C- \\ | & | \end{array}$$

with the understanding that this can be repeated indefinitely. Where compounds are formed with two or more atoms of the same element directly linked, they are usually unstable, as hydrogen peroxide, H_2O_2:

$$H-O-O-H$$

Carbon does have other unique features, but none so obvious as this.

It is now at once possible to see why organic compounds can be so complex: they may contain many hundreds of carbon (and other) atoms. Moreover, if we have a large number of atoms to arrange in molecules, there must be a very great number of ways of doing so, just as the number of possible recipes will increase rapidly as one adds to one's stock of ingredients. Addition of just one new material will open up many more interesting possibilities. Inclusion in a formula of only one extra carbon atom may similarly give rise to many different compounds. It is, in fact, possible to have one molecular formula doing duty for several (or many) different substances. This is because the molecular formula tells us only *how many* atoms of each kind are present in the molecule; it gives no information as to how they are arranged in it, and there may well be more than one possible way. This is the phenomenon of isomerism:

Isomerism – the existence of two or more compounds having the same molecular formula.

Compounds showing this are isomers:

Isomers—compounds having the same molecular formula (i.e., the same numbers of the same atoms), but having their constituent atoms differently arranged.

An example of isomerism is that shown by the two butanes, present in 'calor gas'; these are both C_4H_{10}, but there are two different ways of arranging these 14 atoms, taking into account the normal rules of valency; hence we should expect to find two different substances corresponding to these two structures, and this indeed happens. The two compounds are rather alike in most respects, but do differ in boiling-points, for example. To show their detailed arrangements, we resort to structural formulae which tell us much more than molecular formulae. They are:

Name†	normal butane	iso-butane
Boiling-point	$-0.5°C$	$-11.7°C$
Molecular formula	C_4H_{10}	C_4H_{10}

Structural formula:

```
         H H H H                H H H
         | | | |                | | |
     H—C—C—C—C—H            H—C—C—C—H
         | | | |                | | |
         H H H H                H | H
                                H—C—H
                                  |
                                  H
```

It is fortunate that organic compounds fall into groups or families, so that numbers can be studied together. Indeed, members of the same chemical family are often astonishingly alike in their properties and can be formed in the same kinds of ways. To illustrate some further basic ideas of organic chemistry, we shall introduce a few of these groups.

ALKANES

These are the simplest compounds in organic chemistry and were formerly called 'paraffins'. They contain hydrogen and carbon only and have their carbon atoms arranged in chains, each atom being linked to the next by a single valency bond. The butanes mentioned above are alkanes. Some other examples are:
Methane, CH_4, the chief constituent of the 'natural gas' obtained from the Sahara and under the North Sea and a most important product of petroleum refining; it is the simplest paraffin;
Iso-*octane* (or *2,2,4-trimethyl pentane*), a constituent of high-grade

---

† Modern names are respectively butane and 2-methyl propane

petrol:

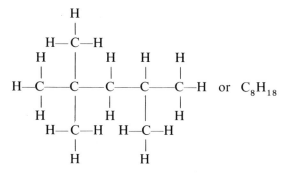

Normal octadecane, present in paraffin wax:

$$\begin{array}{c}
\text{H H H H H H H H H H H H H H H H H H} \\
\mid\ \mid\ \mid\ \mid\ \mid\ \mid\ \mid\ \mid\ \mid\ \mid\ \mid\ \mid\ \mid\ \mid\ \mid\ \mid\ \mid\ \mid \\
\text{H–C–C–C–C–C–C–C–C–C–C–C–C–C–C–C–C–C–C–H} \\
\mid\ \mid\ \mid\ \mid\ \mid\ \mid\ \mid\ \mid\ \mid\ \mid\ \mid\ \mid\ \mid\ \mid\ \mid\ \mid\ \mid\ \mid \\
\text{H H H H H H H H H H H H H H H H H H}
\end{array}$$

ALKENES

These are also hydrocarbons, and their carbon atoms are also arranged in chains. They differ from the alkanes in having one or more pairs of carbon atoms linked by a double bond. Unlike the alkanes, they are chemically very reactive. Examples are: *Ethylene*, made from petroleum gases and converted to polyethylene by high pressures and/or temperatures:

$$\begin{array}{c}
\text{H} \quad\quad \text{H} \\
\diagdown \quad \diagup \\
\text{C}=\text{C} \\
\diagup \quad \diagdown \\
\text{H} \quad\quad \text{H}
\end{array}$$

Butadiene, used in the manufacture of synthetic rubber:

$$\begin{array}{c}
\text{H} \quad\quad \text{H} \quad\ \text{H} \quad\quad \text{H} \\
\diagdown \quad\ \ \mid \quad\ \mid \quad\ \diagup \\
\text{C}=\text{C}—\text{C}=\text{C} \\
\diagup \quad\quad\quad\quad\quad\quad \diagdown \\
\text{H} \quad\quad\quad\quad\quad\quad \text{H}
\end{array}$$

CYCLIC COMPOUNDS

These have the carbon atoms arranged not in an open chain but in a ring, often of six members. They, in fact, include many series. The following hydrocarbons are very important:

cyclo-hexane   benzene

ALCOHOLS

These contain oxygen in addition to hydrogen and carbon. The oxygen is attached by its two valencies to an atom of hydrogen and to one of carbon:

$$-\text{C}-\text{O}-\text{H}$$

Thus all alcohols contain this hydroxyl group, —OH. Several are of great importance to bakery.

ETHANOL
This is the chemical name for the ordinary 'alcohol' present in beverages such as beer, whisky, etc. It can be written:

C_2H_6O or   H—C—C—O—H   or CH_3CH_2OH

Molecular   Structural formulae
formula

It is manufactured by fermentation of carbohydrates (p. 134), or by adding water to ethylene in the presence of a catalyst:

$$C_2H_4 + H_2O = CH_3CH_2OH$$

Ethanol is a powerful solvent and mixes with water in any proportion. Its boiling point is 78°C (= 172°F). It is moderately poisonous and affects first the higher centres of the brain which it tends to dull. For centuries it has been consumed in fermented beverages and in small quantities seems to be free from undesirable effects. It is, of course, formed in baking with yeast, but much is lost in the process, owing to its volatility at the temperatures involved. Oxidation to acetic acid is important (p. 174).

The following beverages are sometimes added in the making of cakes and puddings; the figures indicate the approximate percentages of alcohol in each: brandy (40); rum (40); sherry (16); ale (6).

OTHER MONOHYDRIC ALCOHOLS

There are many other monohydric alcohols (i.e. those with only one OH group per molecule). They include *methanol*, CH_3OH, a poisonous liquid obtained from wood or petrochemicals and used as a fuel, solvent, chemical intermediate and additive to ethanol in methylated spirit to make it (theoretically) undrinkable. There is also *pentanol* or *amyl alcohol*, $C_5H_{11}OH$, found in oven gases and proposed as a measure of aroma intensity; it is a by-product of fermentation.

GLYCOLS

Many important alcohols are polyhydric, i.e., having more than one OH group. The simplest example, *ethylene glycol*, is the 'antifreeze' of car radiators, but, being quite poisonous, is of no concern in food technology. *Propylene glycol*, or *propane-1,2-diol*, is made by hydrating propylene oxide (itself a product of oxidation of propylene and a bactericide for flour):

$$CH_3.CH=CH_2 \longrightarrow CH_3.\underset{\underset{O}{\diagdown\diagup}}{CH-CH_2} \xrightarrow{H_2O} CH_3.\underset{\underset{OH\ \ OH}{|\ \ \ \ |}}{CH-CH_2}$$

propylene   propylene oxide   propylene glycol

Like all alcohols it combines with acids to form esters. These have a bland taste and are used as emulsifiers in batter, sponge cakes and whipped toppings.

GLYCEROL, OR GLYCERINE

This alcohol has three hydroxyl groups in each molecule, and has the structure

$$H-\underset{\underset{OH}{|}}{\overset{\overset{H}{|}}{C}}-\underset{\underset{OH}{|}}{\overset{\overset{H}{|}}{C}}-\underset{\underset{OH}{|}}{\overset{\overset{H}{|}}{C}}-H$$

It is produced when fats are hydrolysed (p. 212), but this time-honoured method is being rapidly superseded by a series of reactions from propene:

$$\mathrm{H-\underset{\underset{H}{|}}{\overset{\overset{H}{|}}{C}}-\overset{\overset{H}{|}}{C}=C\genfrac{}{}{0pt}{}{\diagup H}{\diagdown H}}$$

which is obtained from treatment of petroleum.

It has a high boiling-point and a sweet taste, and dissolves in water in all proportions. Because it is non-poisonous and quite pleasant-tasting, it is often added to medicinal and other preparations to keep them moist in conditions under which water would evaporate. It is used as a lubricant for watches, and reacts with nitric and sulphuric acids to form glycerol trinitrate which is the basis of the explosive 'dynamite'.

For organic acids see p. 174.

Chapter 22

Chemical Basis of Food

The human body is very like a machine. As such, it has certain requirements if it is to function efficiently. Most obvious of all is fuel, but it also needs repair, and these are both given by an adequate diet. Broadly we may say that carbohydrates and fats supply the body's need of fuel, while proteins enable it to effect its own 'running repairs', replacing dead cells by new ones, a process that continually occurs. In addition to these three items there is a fourth which does not have a function so closely paralleled in machinery. This is a group of compounds, known as vitamins, which are essential to healthy growth and living, but which are needed in much smaller quantities. The reason why it is not so easy to supply a mechanical analogy to these is that a machine has no growth nor is liable to disease, and vitamins promote the former and help to stem the latter.

It may be mentioned that, of course, the body also requires water in large amounts together with mineral salts as sodium chloride, phosphates, calcium, etc. As the chemistry of these has been discussed earlier (Chapters 17, 18 and 19), we omit further reference to these, and turn now to deal with the four main organic classes of compounds necessary to diet.

CARBOHYDRATES

These were once defined as naturally occurring compounds of formula $C_x(H_2O)_y$. This is not completely satisfactory in some ways, but will be enough for our purposes now. The chief members of the series are:

Starch: $(C_6H_{10}O_5)_n$, where n is several hundred;
Cellulose: $(C_6H_{10}O_5)_n$, where n is of the order of 1000 or more;
Sucrose, or cane-sugar, or beet-sugar: $C_{12}H_{22}O_{11}$;
Lactose, or milk-sugar: $C_{12}H_{22}O_{11}$;
Maltose: $C_{12}H_{22}O_{11}$;

Glucose: $C_6H_{12}O_6$;
Fructose, or fruit-sugar: $C_6H_{12}O_6$.

It will be noted that glucose and fructose have the same molecular formula, also sucrose, maltose and lactose; they therefore show isomerism, and indeed this phenomenon is fundamentally important in carbohydrate chemistry.

The first two carbohydrates mentioned above are known as 'polysaccharides'; they are insoluble in water. The remaining five are sugars and are all soluble in water and sweet to the taste. In addition, they have a property known as 'optical activity'. This is demonstrated when their solutions are subjected to the action of plane polarised light, i.e., light whose vibrations are all confined to one plane. If this is passed through a solution of an optically active material, the plane of polarisation is rotated through a definite angle, depending on the nature of the substance present, its concentration, the length of the solution through which the light passes, the temperature, and the wavelength of the light.

For example, if plane polarised light is passed through a solution of glucose, sucrose or lactose, its plane of polarisation will be turned to the right, but with fructose it is rotated to the left. An instrument in which these changes can be observed is known as a polarimeter.

1. Glucose

This occurs widely in plants, and is one of the first carbohydrates formed by photosynthesis (p. 136). In many plants glucose molecules become linked together, with elimination of water, to form more complex products, including starch and cellulose. It is from these that glucose is chiefly made industrially; they are decomposed by dilute hydrocholoric or sulphuric acids on heating and often under pressure. The acids catalyse the reaction:

$$(C_6H_{10}O_5)_n + nH_2O = nC_6H_{12}O_6$$

Glucose is a white, soluble solid which is fairly sweet and is directly absorbed by the body into the bloodstream. This is not true of the other common carbohydrates and explains why glucose is so favoured as an ingredient of invalid diets and as a component of energy-producing sweets for children and athletes.

Chemically, glucose is a fairly strong reducing agent. It will, for example, reduce to metallic silver a solution of silver nitrate that has been treated with ammonia. This has been used in the silvering of mirrors, as the silver is deposited as a thin even film on to clean glass surfaces. It also serves to distinguish glucose from sucrose which cannot do this.

CHEMICAL BASIS OF FOOD 209

Glucose is fermented to alcohol and carbon dioxide under the catalytic influence of the enzyme zymase present in yeast:

$$C_6H_{12}O_6 = 2CO_2 + 2C_2H_5OH$$

This occurs when yeast is used in baking† and in the manufacture of alcohol, as glucose is produced by breakdown of more complex carbohydrates (sucrose, starch, etc.) in the first stages of these processes. With the increased availability of glucose syrups having high fermentable content, glucose baking is fast expanding.

It may be estimated by measurements with the polarimeter, by determining the volume of carbon dioxide produced in fermentation or by chemical methods involving the estimate of the reducing power of glucose specimens.

2. FRUCTOSE

This is also present in many fruits. It is much sweeter than either glucose or sucrose, and also differs from them in rotating the plane of polarisation of polarised light to the left. Like glucose, it is a reducing agent, and it can be fermented to alcohol and carbon dioxide, though rather less easily. It is used in diabetic preparations, as in diabetes it is assimilated, unlike other common sugars.

3. SUCROSE

More of this substance is produced commercially each year than of any other pure organic chemical. It is extracted from sugar-cane or sugar-beet by breaking up the plant into small pieces, extracting the sugar as its solution in water and subjecting it to a series of purifying processes, including filtration through charcoal beds which tend to remove coloured impurities. Brown sugar has been less completely refined than white.

Sucrose is a white crystalline solid which is sweeter than glucose but less sweet than fructose. It is the only common sugar that has no reducing properties, and this serves as a most useful test for it. The most important reaction it shows is probably its inversion. If sucrose is boiled with dilute acids or allowed to stand in contact with water and the enzyme invertase (present in yeast and in the alimentary canal) it is converted into glucose and fructose:

$$\underset{\text{sucrose}}{C_{12}H_{22}O_{11}} + H_2O = \underset{\text{glucose}}{C_6H_{12}O_6} + \underset{\text{fructose}}{C_6H_{12}O_6}$$

Of these three sugars, sucrose and glucose rotate the plane of

---

†In the CBP, where the dough is aerated by high speed mixing the stage of bulk-fermentation is of course eliminated.

polarisation of polarised light to the right, but fructose turns it strongly to the left, so much so, in fact, that the mixture of fructose and glucose shows an overall rotation to the left. Thus the sign of rotation has been inverted (from right to left), and the mixture of glucose and fructose is known as 'invert sugar'. This mixture is found in many natural sources and results from this reaction. Owing to the excessive sweetness of fructose, invert sugar is a good deal sweeter than cane-sugar. This accounts for the sweetness of honey, 'golden syrup', and other preparations in which it occurs.

Sucrose is incapable of fermentation by zymase. But as yeast contains invertase as well as zymase it will induce first inversion and then fermentation. Hence sugar solutions will not keep for long in ordinary air, as this is liable to contain minute spores of yeast.

Sucrose is the most important sugar present in flour, though even then it is only there to the extent of about 1.5–2%. It is chiefly on this sugar that the yeast acts when it is added to cause fermentation.

4. MALTOSE

This is a degradation product of starch (see below). It is converted by water into glucose only, either by acids or by the action of the enzyme maltase which is present in malt. It is a reducing agent, unlike sucrose.

A little maltose is usually present in flour, though not often more than 0.5%. Excess of this causes the gluten to soften and the crumb to be gummy. When yeast is used in baking, the maltose also is destroyed.

5. LACTOSE

This sugar is present in the milk of mammals. It is not very sweet, and can be converted by acids and certain enzymes into glucose and the isomeric galactose:

$$\underset{\text{lactose}}{C_{12}H_{22}O_{11}} + H_2O = \underset{\text{glucose}}{C_6H_{12}O_6} + \underset{\text{galactose}}{C_6H_{12}O_6}$$

6. STARCH

Starch is one of the ultimate products of photosynthesis and is stored in plants as a reserve material for producing energy. It is particularly often found in tubers and seeds. Flour contains nearly 70% by weight of starch, and it is also an important constituent of rice (75–80%) and potatoes (20%).

As an item of diet, starch is well known both as a source of energy and also as a contributor to the stores of body-fat. In both these aspects of its action it behaves like other carbohydrates, and it is in

CHEMICAL BASIS OF FOOD

fact broken down in the body into simpler members of the series. The process is complex and takes place in many steps, the most important of which are these, the first being catalysed by diastase:

$$\text{starch} \longrightarrow \text{dextrins} \longrightarrow \text{maltose} \longrightarrow \text{glucose}$$

The dextrins are gummy materials rather like starch except that they are water-soluble; they are used in preparation of some adhesives.

The degradation to maltose occurs in the body under the influence of enzymes secreted in the saliva and pancreatic juices. The maltose is converted to glucose in the intestines by the enzyme maltase. Thus starch like the other carbohydrates mentioned above, is eventually changed into glucose, which is the form in which the body can deal with carbohydrates most easily.

Although the value of n in the formula $(C_6H_{10}O_5)_n$ is very high, starch has a fundamentally simple structure. It is built up of glucose units arranged in long chains. There are two main constituents: amylose (in which the chains are unbranched though coiled) and amylopectin (in which they are linked in tree-like clusters). The products of the decomposition by water (hydrolysis) are simply shorter and shorter chains or 'stumpier' clusters, until at last we are left with just glucose itself. The hydrolysis is analogous to the cutting into smaller lengths of long rows of identical beads.

Starch itself is not strictly soluble in water, though when heated with water, it starts to swell (60°C) and by 70°C forms a good paste of a gelatinous character. This fact is used in laundering (starching) and occurs to some extent during baking. Its role in this process is complex but appears to be essential, being connected with its great ability to absorb water. Added starch phosphates improve a loaf's appearance and volume, and other starch derivatives (esters) are used as thickeners for pie and cake fillings.

A sensitive test for starch is the action of iodine. If this element is dissolved in ethanol or in potassium iodide solution it gives a dense blue colour to the starch. This is due to the amylose. The colour is destroyed by heating but returns on cooling. Dextrins may give a reddish colour under these conditions.

7. CELLULOSE

This carbohydrate is the chief constituent of the walls of plant cells. Unlike starch its function is structural. With acids it can be hydrolysed into, ultimately, glucose only, so we may rightly infer that like starch it is built up of glucose units, arranged in chains of great length. But there is a vital difference between starch and cellulose: cellulose is not in ordinary circumstances converted into glucose in the human body. It is, of course, digested by herbiverous animals, but not normally by man. His enzymes can act on starch but not on cellulose, as the glucose units are linked in different ways. The analogy has been suggested of a zip-fastener, which can only be undone from one end, and where the zip must be the right way round; enzymes which can 'unzip' starch cannot attack cellulose.

These facts are important because cellulose is found in many foodstuffs, and its effect here is merely to add 'bulk', not nourishment. This is not necessarily a bad thing, however, as the passage of undigested cellulose through the intestine helps to stimulate it mechanically and thus facilitates excretion; but it may impose an extra strain on the digestive system, and may be fermented into methane (CH_4) and carbon dioxide. Moreover, cellulose membranes enclosing the cells of foodstuffs may prevent access of the digestive juices to the starch and protein within the cells, but this difficulty is overcome by cooking when the cellulose envelopes burst, and this is one of the most fundamental advantages of cooked food.

White flour contains little cellulose (1.5–2%) but increasing amounts are present in whole flour and in brown flour.

Revision 22.1

1. What is 'invert sugar'?
2. How could you distinguish between the following: sucrose, glucose, maltose, starch and cellulose?
3. How are sucrose and glucose made commercially?
4. Which carbohydrates are present in flour?

FATS

Fats, like carbohydrates, are used by the body as fuel. They are, however, more efficient fuels, as the same amount of heat as that produced by one pound of carbohydrate may be obtained from less than half that weight of fat. It is probably possible to obtain all one's energy requirements from carbohydrates alone but the quantity that would have to be consumed would make the process inconvenient, to say the least. However, they are more quickly digested than are fats, so a carbohydrate diet supplemented by fats is usual. It must be noted that body fat can be formed both from fats ingested as part of the diet and from carbohydrates which the body transforms into fats.

The chemical nature of fats is not, as with carbohydrates, summarisable in a simple molecular formula. They may best be defined as compounds formed from glycerol (p. 205) and organic acids; the

CHEMICAL BASIS OF FOOD 213

latter may be written as

$$C_xH_y-C\begin{matrix}\nearrow O \\ \searrow O-H\end{matrix}\quad \text{or more simply as } C_xH_yCOOH$$

The two components of fats are joined in such a way that if the molecule is broken down by water they will be liberated; chemists call such a combination an 'ester', i.e., the product from combination of an alcohol and an acid with elimination of water.

So far as the body is concerned, the fats chiefly important have x as an odd number, often 15 or 17. An example is the fat known as 'stearin' derived from stearic acid, $C_{17}H_{35}COOH$. The structure of this, together with the products of hydrolysis, can be represented thus:

$$\begin{matrix} C_{17}H_{35}COO.CH_2 & & & & HOCH_2 \\ | & & & & | \\ C_{17}H_{35}COO.CH & + 3H_2O = 3C_{17}H_{35}COOH + & HOCH \\ | & & & & | \\ C_{17}H_{35}COO.CH_2 & & & & HOCH_2 \\ \text{stearin} & \text{water} & \text{stearic acid} & \text{glycerol} \end{matrix}$$

Other important acids giving rise to fats include:

$C_{17}H_{33}COOH$ Oleic acid   ⎫   Present, with stearic acid, in
$C_{15}H_{31}COOH$ Palmitic acid ⎬   most animal fats and many
                                                    ⎭   vegetable oils.
C_3H_7COOH Butyric acid   Present in butter, with many others.

It is often true that a given fat produces on hydrolysis not three molecules of the same acid but three different molecules, or one of one acid and two of another.

Sources of fats are of two kinds, animal and vegetable. Lard, mutton and beef fat, and butter are obviously animal in origin and are solids that fail to dissolve in water but are readily soluble in chloroform, ether, and hot alcohol. A hundred years ago these were all that were available; cakes often had a coarse, open grain (due to large crystals of lard). From plants, on the other hand, are derived fats that are normally liquid at room temperature, though with similar solubilities to the solid fats; olive oil is the commonest example and, in some countries, is preferred to animal fats for cooking purposes.

These liquid fats generally give rise to oleic or related acids. Since the early 20th century use has been made of their ability to take up hydrogen to form other acids; thus

$$C_{17}H_{33}COOH + H_2 = C_{17}H_{35}COOH$$
oleic acid   stearic acid

When this reaction is performed on the fats themselves, the products tend to be solid instead of liquid, and this is the basis for the manufacture of margarine in which the oils are reacted with hydrogen in the presence of a catalyst of nickel. The product so obtained is relatively deficient in several vitamins (p. 217) as compared with animal fats, but since 1940 it has been statutory for manufacturers to add specified amounts of vitamins A and D. Margarine was originally introduced as a means of supplementing available supplies of animal fat which were not sufficient owing to the wartime necessity 'to convert as much fat as possible into glycerol so as to make explosives (p. 206). Some margarine is now made from whale-oil—an animal fat but a liquid.

As mentioned above, fats have an importance quite apart from their value as food, and this is their hydrolysis to glycerol. This reaction occurs in the body (when fats are broken down under the influence of fat-splitting enzymes as lipase, and then resynthesised). It can also be brought about by heating with superheated water, i.e., water that has been heated under pressure to temperatures above 100°C. Enzymes can be used on the large industrial scale as well as in the laboratory. The chief method of hydrolysis is that of heating with alkali. When this is done the products are glycerol and the alkali salt of the acids, so the reaction becomes (for caustic soda):

$$\begin{array}{l} C_xH_yCOO.CH_2 \\ | \\ C_xH_yCOO.CH + 3NaOH = 3C_xH_yCOONa + \\ | \\ C_xH_yCOO.CH_2 \end{array} \quad \begin{array}{l} HOCH_2 \\ | \\ HOCH \\ | \\ HOCH_2 \end{array}$$

Where x is large, as it usually is, the sodium salt is in fact soap†, which can be defined as the sodium or potassium salt of a long-chain fatty acid. It exerts its detergent action because part of the molecule, that containing the —COONa group, has an 'affinity' for water, while the very long chain of carbon atoms tends to be soluble in oils and grease. Thus the water and oil are 'brought together' and removal of one by the other becomes easier. All detergents have this action, though most do not suffer from the disadvantage of soap itself of forming insoluble calcium and magnesium salts with hard

---

†Hence this reaction with alkalis is termed 'saponification' i.e. 'soap-formation'

CHEMICAL BASIS OF FOOD

water. It is this which has done much to displace soap as the standard detergent.

Fats are combinations of acids and glycerol; it does not follow, however, that there are always three molecules of acid linked to every one of glycerol. It is possible, for instance, to have compounds of the type:

$$\begin{array}{c} C_xH_yCOO.CH_2 \\ | \\ HOCH \\ | \\ HOCH_2 \end{array} \quad \text{e.g.} \quad \begin{array}{c} C_{17}H_{35}COO.CH_2 \\ | \\ HOCH \\ | \\ HOCH_2 \\ \text{(monostearin)} \end{array}$$

These monoglycerides are often pleasant-tasting, non-volatile liquids, and they are being used as additives in bread-making for they help to keep the loaf moist, especially important in sliced bread. Such substances, often called emulsifiers, have been increasingly incorporated into baking mixtures since the 1930s. Their use has been justified by the fact that they are formed from fats in the body (intestines) during their transformation to glycerol:

$$\text{fats} \longrightarrow \text{monostearin} \longrightarrow \text{glycerol}$$

In general fat systems play a variety of roles in the baking process. They impart flavour, richness and tenderness. They promote a desirable texture, make possible the development of flakiness in puff pastry, and encourage aeration and consequent leavening. Much of this arises from the emulsifying action of some of the fat components.

Revision 22.2
1. Compare carbohydrates and fats as foodstuffs.
2. Define the term 'fat'.
3. What useful products are obtained from fats?

PROTEINS

Carbohydrates and fats are both utilised by the body as fuels to produce energy. This could be true of proteins, but they have a function which the other constituents of diet do not. Proteins are the raw materials for the replacement of body cells, and are the chief components of living tissue (excluding bone). In addition, they increase the production of body-heat.

Proteins are all complex molecules containing nitrogen. Like starch and cellulose they are made up of a large number of smaller units strung together like beads on a necklace. But unlike these carbohydrates, proteins are made of different units and they differ from each other in (a) the nature of the units, (b) their number, and (c) the order in which they are arranged. These units are known as amino-acids, and are of the form:

$$H_2N-CH-COOH$$
$$|$$
$$R$$

where R can be a wide variety of groups. One such amino-acid, L-cysteine (where $R = HS.CH_2 -$), is now met in baking practice as a reducing agent used as a flour additive and sponge-mixture component. The amino-acids are linked together by what are called 'peptide linkages',

$$-N-C- \quad \text{or} \quad -NH.CO-$$
$$| \quad \|$$
$$H \quad O$$

the nitrogen atom of one being joined to the carbon atom of the next.

Like fats and carbohydrates they are hydrolysed by enzymes in the body to simpler materials, ultimately forming their constituent amino-acids. Thus suppose we have a protein† made of six amino-acid molecules as follows, where R, R′ and R″ are all different; it would be broken down by hydrolysis in this way:

$$\text{HOOC.CH.NH.CO.CH.NH.CO.CH.NH.CO.CH.NH.CO.CH.NH.CO.CH.NH}_2$$
$$|\quad\quad |\quad\quad |\quad\quad |\quad\quad |\quad\quad |$$
$$R \quad\quad R \quad\quad R'' \quad\quad R' \quad\quad R \quad\quad R''$$
$$\downarrow + 5H_2O$$
$$2\text{HOOC.CH.NH}_2 \quad + \quad 2\text{HOOC.CH.NH}_2 \quad + \quad 2\text{HOOC.CH.NH}_2$$
$$|\quad\quad\quad\quad\quad\quad |\quad\quad\quad\quad\quad\quad |$$
$$R \quad\quad\quad\quad\quad\quad R' \quad\quad\quad\quad\quad\quad R''$$

This happens in the body, and resynthesis into other proteins then occurs. It has been shown that certain amino-acids must be present in a balanced diet, so it is more important to know what a protein gives on hydrolysis than where it comes from, or even the order in which the amino-acid units are arranged.

From the point of view of diet, animal proteins have been definitely shown to be superior to those from vegetable sources as they

† This 'protein' is completely hypothetical, because so simple a compound would be technically termed a 'peptide'; but the principle of the hydrolysis is quite correctly shown.

contain more of the essential amino-acids. But some animal proteins, e.g. gelatin, lack altogether some of these, so it is obviously wise to have as wide a variety of proteins as possible.

Animal sources of protein vary in the amount they contain from chicken breast and cheese (25%), through lean meat and fish (roughly 20%), to eggs (12%), and milk (3%). Vegetable sources include soya-beans and peanuts (around 30%), and flour (12%). The last-named substance is deficient in some essential amino-acids (and also in phosphorus), and has two fairly simple proteins, gliadin and glutenin. Despite its deficiencies flour is valuable as the cheapest source of protein. White flour has a little less protein than whole flour. One advantage of the CBP is that it can operate successfully with flours of lower protein content than would lead to acceptable products by the conventional baking process.

Proteins vary widely in their properties. Some, e.g., gliadin, are soluble in alcohol, while others, e.g., glutenin, are insoluble. Few will dissolve in water to give true solutions, but they all show colloidal behaviour (p. 99). Heat and treatment with acids often induce physical changes associated with uncoiling of protein spirals, and is known as 'denaturation'. This occurs in baking.

Revision 22.3
1. What function have proteins in the diet?
2. Comment on bread as a source of protein.

VITAMINS

A diet of fats, carbohydrates and proteins will not be adequate unless it is supplemented by the addition of small amounts of vitamins. These are organic compounds of which small quantities are needed in diet for the maintenance of health. All have been discovered this century, most occur in plants and many have been made synthetically. A few of the more important are mentioned below.

1. VITAMIN A

The yellow pigment in carrots and tomatoes is converted in the body to a closely-related compound that is essential for growth in young animals, and whose deficiency causes 'night-blindness'. This substance is vitamin A. It is chiefly obtained from the livers of cod or halibut in which it is stored in great quantities. It is also present in the milk of cows that have fed on green food (and, therefore, in butter), and occurs in the germ of wheat.

It is destroyed on heating in air, and for this reason may be lost by poor cooking. It is soluble in fats but not in water, and is an alcohol of formula $C_{20}H_{29}OH$. When flour is bleached by chlorine dioxide, vitamin A and related substances are oxidised. In other circumstances, vitamin A is a standard additive to flour.

2. Vitamin B
This name covers a range of vitamins known as B_1, B_2, B_3, etc. Deficiency of these causes various diseases from nervous disorders to pernicious anaemia. Vitamin B_1 is found in the bran layers surrounding the wheat germ, and its presence there constitutes one of the chief advantages of wholemeal over white. Other members of this group are present in flour and may nearly all be found in yeast. One of these compounds, nicotinic acid, or niacin, is now added to flour.

3. Vitamin C
Lack of fresh greens has long been known to cause scurvy. We now know that the compound called vitamin C is present in greens, and in higher concentrations in citrus fruits, rose-hips and especially black currants. Inclusion of this in the diet produces resistance to scurvy and (in fairly large doses) to disease generally. It is soluble in water, tends to be destroyed by heating and is a strong reducing agent. Its formula is $C_6H_8O_6$. It is not present in wheat, but is usually added as an improver to flour (except wholemeal).

4. Vitamin D
This is also added to flour as an improver. Lack of it halts the development of bones and teeth and gives rise to rickets. Its is formed in the body on exposure to sunlight, but can be obtained in addition from cod-liver oil.

5. Vitamin E
This vitamin is found in wheat germ and other places, and is usually extracted from an oil obtained from sprouting wheat germs. It is an anti-sterility vitamin, and apart from being necessary in food is sometimes used medicinally.

Revision 22.4
1. What is a vitamin?
2. What vitamins would you expect to find in a sample of flour, and what beneficial effects do they have?

Appendix: Atomic Weights

The table gives the approximate atomic weights of some of the more important elements.

Element	Symbol	Approximate Atomic Weight
Aluminium	Al	27
Antimony	Sb	122
Argon	A	40
Barium	Ba	137
Bromine	Br	80
Calcium	Ca	40
Carbon	C	12
Chlorine	Cl	35.5
Chromium	Cr	52
Copper	Cu	63.5
Gold	Au	197
Hydrogen	H	1
Iodine	I	127
Iron	Fe	56
Lead	Pb	207
Magnesium	Mg	24
Manganese	Mn	55
Mercury	Hg	200
Nickel	Ni	59
Nitrogen	N	14
Oxygen	O	16
Phosphorus	P	31
Potassium	K	39
Silver	Ag	108
Sodium	Na	23
Sulphur	S	32
Tin	Sn	119
Zinc	Zn	65

Answers to Numerical Exercises

Revision 1.1   p. 5
3. 60 000 J
4. 10 W (or 0.01 kW)
5. 18 000 J

Revision 1.2   p. 7
2. milk, water, new laid egg

Revision 2.1   p. 10
2. large coal, shelled walnuts

Revision 3.1   p. 18
1. 75 cm, 25 g
3. 77 000 kg/cm^2
4. 10.01 kg

Revision 3.2   p. 20
6. 4704 N/m^2
7. 100 cm^3, 600 cm^3

Revision 4.2   p. 28
2. 0.14 N
5. 0.01 mm

Revision 7   p. 46
2. (a) 4200 J
   (b) 2.1 × 10^7 J
   (c) 2100 J
   (d) 360 000 J
3. 5°C
4. 125 g
5. 3600 J/kg°C
6. 5.8°C

Revision 11   p. 85
4. 960 ohms, 3p
6. (a) 0.2 A, 0.13 A
   (b) 0.33 A
   (c) 6 ohms
8. 768 ohms, 576 ohms, 137 V, 103 V, 42.8 W
10. 13A, 13A, 3A

ANSWERS TO NUMERICAL EXERCISES   221

Revision 18.1   p. 164   3. 5.0, 4.75, 10.43

Revision 18.2   p. 171   3. 88, 90, 122, 75, 41, 32.7
                             4. 0.28M, 0.445M, 0.25M, 0.515M
                             5. 95.4, 18.30, 31.2, 105.8

Revision 20.1   p. 190   2. Cl:O
                                Ca: $+2$; F: -1
                                H: $+1$; S: -2
                             3. H: 0 to $+1$
                                S: $+4$ to $+6$
                                N: -3 to $+3$
                                None
                                H: 0 to $+1$
                                C: 4 to 0
                                Zn: 0 to $+2$

Index

Acetates, 175
Acetone, 98, 195
Acetylene, 130, 155, 195
Acid
  Acetic, 174–5
  Butyric, 213
  Carbonic, 136, 155
  Citric, 176
  Hydrochloric, 173
  Nitric, 147, 155, 173–4
  Nitrous, 147, 155
  Oleic, 213
  Palmitic, 213
  Phosphoric, 155, 181–2
  Stearic, 213–14
  Sulphuric, 155, 172–3, 182
  Sulphurous, 155
  Tartaric, 172, 173, 175, 182, 184
Acids
  basicity, 162
  characteristics, 160
  definition, 162
  examples, 161, 172–6
  formation, 155
  ionisation, 162
Air
  composition, 122–6
  expansion of, 30
  fractionation, 125, 126, 131, 138
  humidity, 140
  pollution, 139
Alcohols, 204–6
Alginates, 102
Alkalis
  characteristics, 160–1

Alkalis—*contd.*
  definition, 162
  examples, 176–8
  formation, 155
Alkanes, 202–3
Alkenes, 203
Alloys, 191
Alternating voltage, 83
Aluminium, 192–3
Amino-acids, 216
Ammeter, 78
Ammonia, 145, 146, 148–9, 153
Ammonium
  bicarbonate and carbonate, 179–80, 184
  hydroxide, 177–8
  persulphate, 198
Amperes, 78
Amylase, 95, 211
Anthracite, 196
Aqua Regia, 174
Archimedes' principle, 11
Argon
  air, in, 125, 129, 137–8
  uses, 138
Atomic fission, 107
Atoms, 25, 76
  characteristics, 106–8
  electrical nature, 115 *et seq.*
  symbols, 109–12

Baking powders, 183–7
Baking tins, 190–1

INDEX

Bases, 160, 162
  see also Alkalis
Battery, 70
Bell, electric, 81
Benzoyl peroxide, 198
Berzelius, J. J., 109
Bicarbonates, 136, 178–80
Bimetallic strip, 31
  thermometer, 34
  thermostat, 32
Bleaching, 197–8
Boiling, 47, 49, 51
  effect of pressure on, 53
Bomb calorimeter, 45
Burette, 166–7
Butadiene, 203
Butane, 188–9, 194, 201–2

Calcium, 150–1
  carbide, 155
  carbonate, 156, 180–1
  chloride, 156
  hydroxide, 177
  oxide, 177, 180
  phosphate, 182, 184
  sulphate, 156–8, 181
Calorimeter, 42
  Bomb, 45
Capillary action, 27
Carbohydrates, 207–12
Carbon
  cycle, 140–4
  dioxide
    air, in, 125
    chemical properties, 136–7
    from baking powder,
      estimation of, 185–7
    physical properties, 135
    preparation, 133–5
    uses, 137
  tetrachloride, 98
Carbonates, 134–5, 178–81
Catalyst, definition, 94
Cell, 71
Cellulose, 175, 207, 211–12
Celsius scale, 70
Chalk, 142

Charcoal, 196
Chemical change, 29, 89
Chlorine, 178, 197, 198
  dioxide, 198
Chorleywood process, 5, 45
Chromium plating, 191
Clark's process, 158
Coal, 142, 144, 146, 195, 196
  gas, 194
Coke, 196
Colloids, 99 *et seq.*
Colour, 68
Combustion, 112 *et seq.*, 129–30,
  139, 141, 193–6
Compound
  definition, 92
  mixtures, and, 91
Conduction, 57
Convection, 57
  currents, 60
  hot water system, 60
  ventilation, 61
Coral, 142
Corrosion, 190–3
Cream of Tartar, 175, 182, 184
Critical angle, 71
Crystallisation, 98
Current, electric, 77
Cyclic compounds, 204

Dalton, J., 105 *et seq.*
Davy lamp, 56
'Death Valley', 136, 142
Denitrifying bacteria, 146
Density
  bulk, 9
  concept of, 6
  measurement of, 7
Detergents, 104, 157, 176, 214–15
Dew point, 54
Dextrins, 211
Diastase, 95, 211
Diffusion of gases, 20
Direct current, 83
Disperse phase, 99
Dispersion medium, 99
Dynamite, 206

Elasticity, 15–17
Electricity, 76
  cost of, 80
Electrode, 117
Electrolysis, 117
Electrolyte, 117
Electromagnet, 80
Electromagnetic effect, 80
Electromagnetic spectrum, 74
Electromotive force, 77
Electrons in electricity, 76
Electrophoresis, 100–1
Electroplating, 191
Element, definition, 92
Emulsifiers, 104, 215
Emulsions, 103
Enamels, 190
Energy
  heat, 48, 79
  light, 67
  potential and kinetic, 4
Enzymes, 94–5, 214
  see also individual enzymes
Epsom salts, 181
Equations, 118 et seq.
Ethanol, 98, 204–5
Ether, 98
Ethylene, 203–4
Evaporation, 26, 50, 52
  effect of temperature, 51
Expansion, 29

Fats, 104, 166, 176–7, 212–15
Fermentation, 133, 141–2, 157
Fire extinguishers, 137
Flour
  Additives
    Ammonium persulphate, 198
    Bleaches, 197–8
    Calcium carbonate, 181
    Calcium phosphates, 182
    Lime-water, 177
    Magnesium sulphate, 181
    Potassium bromate, 198
    Sodium chloride, 178
    Sodium sulphate, 181
  A mixture, 91–2

Flour—contd.
  Colloidal behaviour, 102
  Components
    Carbohydrates, 210–12
    Proteins, 217
    Vitamins, 217–8
  Ignition, 130
Flow cup, 23
Fluids, 1, 11
Force, definition, 2
Formulae
  molecular, 109–12, 201–2
  structural, 112, 174–6, 202–6
Frequency
  a.c. mains, of, 83
  waves, of, 74
Fructose, 209
Fuels, 193–6
'Fur', 157
Fuses, 82

Galvanising, 191
Gases, 1, 19, 47
Gelatin, 102, 217
Gels, 102–3
Gliadin, 102, 217
Glucose, 133–4, 142, 207–10
Gluten, glutenin, 102, 198, 217
Glycerol, 177, 205–6, 212–15
Glycols, 205
Gravity, force of, 2

Heat, effects of, 29
Hooke's Law, 16
Hot water system, 60
Humidity, 54, 140
Hydration (in colloids), 102
Hydrogen
  preparation, 150–2
  properties, 152–3
  uses, 153
Hydrometer, 12
Hygrometers, 54, 55

Images, 69

INDEX

Indicators, 154, 155, 160, 163–4
Inert gases, 137–8
Ingen-Hauz apparatus, 58
Insulators (electric), 76
Invert sugar, 210
Invertase, 95, 134, 209
Ion(s), 115 *et seq.*, 162–4
  exchange, 158
Isomerism, 201
Isotopes, 108

Joule, 4

Kelvin scale, 34
Kilowatt, 4
Kilowatt-hour, 80

Lactose, 210
Latent heat, 48
Lavoisier, A., 127
Law of Constant Composition, 91
Leslie's cube, 63
Light, 67 *et seq.*
  polarised, 72
  reflection of, 68
  refraction of, 70
Limestone, 142
Lime-water, 136, 177
Lipase, 95, 215
Liquids, 1, 18, 47
Luminous bodies, 67
Lyophilic sols, 102
Lyophobic sols, 101–2

Magnesium
  chloride, 156
  sulphate, 156, 181
Maltase, 95, 210
Maltose, 210
Manganese dioxide, 94, 127
Margarine, 153, 214
Mass, 2
Matter, states of, 1, 47

Maximum and minimum
  thermometer, 38
Mechanical dough development, 5, 45
Mercury (II) oxide, 127
Methane, 194, 202
Method of mixtures, 42
Microwave ovens, 75
Milk, 103–4, 210
Mixture and compounds, 91
  definition, 92–3
Molarity, 165 *et seq.*
Mole, 165
Molecular weight, 111
Molecules, 20, 25
Monoglycerides, 215

Newton, 3
Nitrates, 181
Nitrifying bacteria, 145
Nitro-compounds, 174
Nitrogen
  air, in, 125, 130
  chemical properties, 131–2
  cycle, 144–9
  dioxide, 147, 155, 173, 198
  fixation, 147
  physical properties, 131
  preparation, 131
  trichloride, 198
  uses, 132
Nitrosyl chloride, 174, 197

Ohms, 78
Ohm's Law, 77
Organic chemistry, 200–6
Ovens
  construction of, 64
  convection currents in, 62
  insulation of, 59
  shelves, expansion of, 31
  temperature of, 34
Oxidation, 188–99
Oxygen
  air, in, 125
  chemical properties, 126 *et seq.*

INDEX 227

Oxygen—*contd.*
 physical properties, 129
 preparation, 126–8
 uses, 130
Ozone, 198

Paraffins, 202–3
Penetrometer, 25
Permutit, 159
Petroleum, 142, 194–5
pH, 163–4
Phosphates, 181–2
Photosynthesis, 143
Physical changes, 29
Pipette, 166–7
Plasticity, 18
Polarised light, 72
 Polaroid, 73
Polythene film, 20
Potassium
 bicarbonate and carbonate, 186
 bromate, 198
 chlorate, 127
 hydroxide, 177
 nitrate, 181
 sulphate, 181
Pressure, 19
Pressure cooker, 53
Priestley, J., 127, 137
Producer gas, 195
Propane, 194
Proteins, 145, 215–17
PTFE, 191
Pulfrich refractometer, 72

Radiation, 54, 62
 Leslie's cube, 63
Radicals, 110, 113–14
Radioactivity, 107
Reduction, 177–86, 188–90
Redwood viscometer, 23
Reflection, 69
Refraction, 70
Refractometer, 72
 Pulfrich, 72
Refrigeration, 51

Relative density, 9 *et seq.*
Rennin, 95, 104
Resistors, in series and parallel, 78
Respiration, 122, 141

Saccharometer, 13
Sal volatile, 180
Salts, 160–2, 178–83
Saturation, 97–8
Scheele, C. W., 127
Sherardising, 178–9, 191
Silicones, 190
Soap, 104, 155–7, 176–7, 214
Soda-waters, 137
Sodium
 bicarbonate, 178–80, 183–4
 carbonate, 178–80
 chloride, 178
 hexametaphosphate, 159, 182
 hydroxide, 176–7
 hypochlorite, 197
 nitrate, 181
 phosphates, 181–2
 sulphate, 181
Solids, 1, 15, 47
Solubility, 97 *et seq.*
Solute, 97
Solutions, 97 *et seq.*
Solvent, 97
Specific gravity, 11
Specific heat, 38
 Bakers' rule of thumb, 42
 method of mixtures, 39
Specific heat capacity, 41 *et seq.*
Specific latent heat, 49
Standards, in titrations, 166–8
Starch, 103, 207, 210–11
Steels, 191–2
Sucrose ('sugar'), 91, 134, 196, 209–10
Sulphates, 181
Sulphur dioxide, 139, 155, 172, 196
Supersaturation, 98
Surface tension, 25
 baking tins, and, 28
 capillary action, 27

Symbiotic bacteria, 148

Tartrates, 175, 182–4
Temperature, 29
  chemical effects of, 93–5
  measurement of, 34
Tensile stress, 17
Thermocouple, 37
  cut-out for ovens, 81
Thermometers, 34
  types of, 36–8
Thermostatic control, 32
  ovens, of, 33
Three-phase supply, 84
Tin-plating, 78
Titrations, 164–71
Total internal reflection, 71
Trichlorethylene, 98
Trypsin, 95, 216

Vacuum flask, 64
Valency, 112 et seq.
Vapour, 47
  pressure, 53
Velocity of waves, 74
Ventilation, 61
Vinegar, 174–5
Viscometers, 23
Viscosity, 22 et seq.
Vitamins
  A, 214, 217–18
  B, 218
  C, 218
  D, 214, 218
  E, 218
Voids, 10

Voltmeter, 77
Volts, 77
Volume
  bulk, 9

Washing soda, 158, 180
Water
  chemical properties, 154–5
  gas, 194–5
  hardness of
    causes, 143, 156–7
    effects, 157
    removal, 157–9
    temporary and permanent, 158
  ionisation, 163
  physical properties, 153–4
  solvent, as, 98
Watt, 4
Watt-hour, 5
Watt-second, 5
Wavelength, 74
Waves, 74
Weathering of rock, 143, 146–7, 154, 156
Weight, 2
Work
  against gravity, 4
  definition, 3
Wrapping materials, 20

Yeast, 95, 134, 142, 209–10

Zinc, 191
Zymase, 95, 134, 142, 209